Questionnaires

Practical Hints on How to Avoid Mistakes in
Design and Interpretation

T.L. Brink, Ph.D., M.B.A.

Heuristic Books

Chesterfield Missouri, USA

Questionnaires — Practical Hints on How to Avoid Mistakes in Design and Interpretation

Graphics Credits:
The cover design is by Robert J. Banis, Ph.D., based on a photograph from the United States Census Collection.

First Printing, April 2004
ISBN 1-888725-74-5

Library of Congress Cataloging-in-Publication Data

Brink, T. L. (Terry L.).
  Questionnaires : practical hints on how to avoid mistakes in design and interpretation / T.L. Brink.
      p. cm.
  Includes bibliographical references and index.
  ISBN 1-888725-74-5 (alk. paper)
  1. Questionnaires--Methodology. I. Title.
  HM538.B75 2004
  001.4'33--dc22

                              2004009218

Heuristic Books
PO Box 7151
Chesterfield, MO 63006-7151
(636) 394-4950
heuristicbooks.com

Contents

Section One: Errors in Initial Approach and Conceptualization

Chapter 1: Do we really need to do a questionnaire?

Gut feelings are not science

MISTAKE 1: Stopping after qualitative research, and foregoing the quantitative follow-up.

EXAMPLE: A manager interviewed a couple of workers about what could be done to improve production. He implemented their suggestions, but no real benefits in productivity occurred.

SOLUTION: Qualitative research is much more vulnerable to small and non-representative sampling: perhaps the above workers were not typical of those of the entire firm. Also, qualitative research is more vulnerable to the investigator's own biases: both in soliciting the data and in interpreting those data. Perhaps the manager in question did not really hear all of what the workers were saying, or perhaps his very tone and presence steered the workers onto one path prematurely. Remember: qualitative research only suggests hypotheses, but it takes quantitative research in the form of a survey or an experiment to confirm a hypothesis.

Don't start measuring until you know what you need to measure

> **MISTAKE 2:** *Jumping over qualitative research and going immediately to quantitative research.*

EXAMPLE: You start investigating a morale problem among workers by constructing a questionnaire. You threw out a large number of questions in hopes that a few would tap the workers' pool of frustrations. You end up getting a low return rate, and many of the questionnaires have obscene comments indicating that the workers did not appreciate the phrasing of some of the questions.

EXAMPLE: A flower delivery service develops a questionnaire about which types of arrangements are preferred by the customers. More of the favored arrangements are offered, but sales continue to decrease dramatically. The number of repeat customers, particularly, is going down.

EXAMPLE: A corporation of dentists surveys patients to find out if opening other branch offices will be considered convenient. The patients overwhelmingly say "yes" so other branches are opened. Overhead costs increase, but the number of patients do not: the current patients just redistribute over the new locations.

SOLUTION: Before we drill for oil, we must make sure that we are drilling in the right place. Before we construct a questionnaire to measure certain variables, we must make sure that those are the issues which workers, or customers, or patients consider to be salient. Initial qualitative approaches, and focus groups in particular, are essential to identifying the right variables to investigate. The workers were asked if they were happy with the cafeteria food, when they were really frustrated over their supervision, so that merely reinforced their perception that supervisors do not have a clue as to real frustrations. Asking flower customers about floral arrangements was equivalent to rearranging the deck chairs on the Titanic: the real problem was that deliveries were coming late, after the anniversary or birthday or Valentines Day! The dental office mistakenly assumed that patients were sensitive to geography, until it was noted that some patients left for a competitor who had evening and weekend ours: the patients were willing to drive a few miles in order to have more convenient hours.

> *MISTAKE 3: Trying to get qualitative data with a questionnaire rather than an interview. Qualitative data are often an essential starting point, because formal questionnaires are rarely effective in getting the kind of rich responses that we need.*

EXAMPLE:

What fringe benefits would you like to see?

If this appeared on a written questionnaire with about a half inch of space underneath, it would not get much of a response. Email and online questionnaires would not do much better. Writing an answer is work, and requires more motivation, preparation, and organization for the subject than an oral response.

SOLUTION: An unobtrusive ethnography might be a good starting point. What do the workers in the lunch room gripe about? What kinds of things do they mention enviously when discussing the jobs of family members or friends? Interviews (whether focus groups, face to face, or over the phone) are greatly superior to structured questionnaires because they allow for interaction between examiner and subject. The interviewer can respond to specific statements and draw the subject out further.

Don't overlook the ease of archival data

> *MISTAKE 4: Going to the trouble of developing of questionnaire when the data you seek are already in easily accessible archives.*

EXAMPLE: The sales manager of an automobile dealership wanted to find out the median and distribution of the income of his clientele. He set out a little questionnaire and ballot box close to one of the cars in the showroom. After a week, he looked in the box and was amazed to find less than a dozen filled out forms. He also suspected that some of the respondents had not been honest, and wildly exaggerated their income.

SOLUTION: Most businesses have access to customer records, employee records, and a file of applications of job seekers. Each customer, employee, and job applicant represents a potential research subject. The data in these archives are usually quantified systematically for easy comparison. In this example of the marketing manager, he merely had to ask the finance manager, who had a file of credit applications taken on every customer who submitted an offer to purchase or lease a vehicle (even the ones who paid cash). Since the applicants had to sign their applications, and agree to a credit check, we could assume that their statements of income were probably more valid than those given on a mere questionnaire. Since every applicant had to fill out one of those applications, the sampling was not only larger, but also more representative of the population.

EXAMPLE: A nurse working on an advanced degree began a new job in an extended care geriatric facility. She wanted to find out what proportion of the patients in the facility were confused and disoriented. She found a valid and reliable psychometric scale, the International Version of the Mental Status Questionnaire, but the director of the facility wanted her to obtain informed consent from the patients (or their families) because their participation in the study would necessitate face-to-face interviews.

SOLUTION: The nurse realized that the information she sought could be obtained from the patient records, without requiring a face-to-face interview of the subjects. The nursing staff regularly rated each patient as "lucid and alert" or "confused and disoriented" (on a five point scale) each day. All this researcher had to do was to look at the patient records. For this particular variable (senile confusion) the ratings of nursing staff are considered to be a valid measure. Indeed, many scales such as the International Version of the Mental Status Questionnaire were validated by comparing patients' answers to the individual questions to the staff ratings of the patients' mental status (using the latter as the established standard measure of the variable).

Don't neglect the usefulness of field data

> *MISTAKE 5: Going to the trouble of developing of questionnaire when the data you seek could be more readily measured by simply observing and counting the actions of the subjects unobtrusively.*

EXAMPLE: The manager of the toy department in a large department store wanted to find out if most children came into the store with their mothers or fathers. He left questionnaires and ballot boxes by several entrances, and also one right in the toy department. Over the next week, he noticed that very few forms had been filled out, and most of those which had been indicated that they were not with children on that trip to the store. Yet, he noticed many children who were in the store with their parents. He then asked one woman with her two kids if she had filled out the questionnaire, and she replied that she did not have time to do so when she was with her children, but perhaps on some other occasion when she was alone, she would have time to fill it out.

SOLUTION: All the toy department manager had to do was to look systematically at the customers entering the store (or perhaps just those in the toy department) and measure two variables: the adult configuration and the child configuration.

ONE WOMAN   ONE MAN   SEVERAL MEN   MEN & WOMEN   MANY ADULTS
NO CHILDREN   ONE BOY   ONE GIRL   SEVERAL CHILDREN

He could even have been more precise, estimating the age of each adult and child. Compared to the use of a questionnaire, this kind of field enumeration would have been quicker, obtained a larger sample, obtained a more representative sample, and on some variables, it would have obtained a more valid measure of the variable.

EXAMPLE: A restaurant owner made sure that each customer got a questionnaire with the bill. The questionnaire asked subjects how interested they would be in extended hours of operation: later, earlier, weekends. The customers indicated that they would not be interested. Some even volunteered the explanation that they were not up and about looking for a restaurant at those times, so the management shelved plans to stay open later. A few weeks later, a competitor down the block tried staying open later, and was highly successful.

SOLUTION: A brief field experiment (actually staying open later) would have been a superior source of information. The open restaurant might have been a stimulus that could have appealed to an entirely different clientele than that which was included in the survey.

Chapter 2: How precisely do we need to measure?

Distinguishing constants, variables and operational definitions

DATA are the bits of information that are observed by psychological research. Within the social sciences, the term "data" is regarded as plural. So, we should say, "these data are" instead of "this data is."

We use the term SUBJECTS to describe the people actually observed and from whom data are gathered about the variables in question. Subjects are people whose dependent variables (e.g., performance, attitudes) are being measured. In business related research, the subjects are usually workers (or job applicants) or customers (or potential customers). In clinical related research, the subjects are usually patients.

So, a subject is WHOM is being measured, but WHAT is being measured is known as a constant or variable. A numerical measurement that does not change is called a CONSTANT. Measurements that can change are known as VARIABLES. In other words, a given case's measure on a different variable can differ from that of another case's: one can be present while the other is absent; one case can be high while the other case is low on the same variable.

> SUBJECT: CUSTOMER #15
>
> BACKGROUND VARIABLE: GENDER (he was a male)
>
> BACKGROUND VARIABLE: AGE (he was 47)
>
> BACKGROUND VARIABLE: FAMILY (he has no spouse, no children)
>
> OUTCOME VARIABLE: PURCHASE (he bought a pickup)
>
> SUBJECT: JOB APPLICANT #42
>
> PRE-EMPLOYMENT PREDICTOR VARIABLE: SKILLS TEST
> (he scored high on mechanical ability)
>
> OUTCOME VARIABLE: PERFORMANCE (he was productive on the job)
>
> SUBJECT: PATIENT #12
>
> TREATMENT VARIABLE: She was assigned to the placebo group.
>
> OUTCOME VARIABLE: Her condition did not improve.

> *MISTAKE 6: Not being able to identify each recorded datum as a measure of a specific variable for a specific subject.*

EXAMPLE: One student doing a field count of customers in a store was rather careless in copying down the numbers. When he got home and looked at his notes, he could not remember which numbers were being used to identify the subjects (e.g., #1, #2, #3) and which were being used to measure the variable of how many minutes they waited in line.

SOLUTION: One solution is to develop a data recording sheet which has the same kind of layout as an Excel spreadsheet: each subject has one row, all of the data for that subject are on that row, and no other subject's data are on that row. Each variable is represented by one column, all of the data on that variable are in that column, no other variable's data are in that column. Another approach is to have a separate card for each subject, and a certain place on each card for the data related to a particular variable. When each subject fills out a separate, pre-formatted questionnaire, these guidelines are met.

Use the right scale for precision

What we study are the variables. Variables are defined by their OPERATIONAL DEFINITIONS: by the numbers we employ to measure the variable.

ORGANISM	DEPENDENT VARIABLE	OPERATIONAL DEFINITION
Rat	Performance running a maze	Number of seconds it took the rat to get through the maze
Voter	Attitude about a political candidate	Whom the voter says that she will vote for
Consumer	Decision to purchase a product	Whether or not the consumer purchases the product
Worker	Absenteeism	How many times last year the worker did not show up for a scheduled shift
Patient	Depression	Score on a valid and reliable depression scale

Data are said to be QUANTITATIVE if they are based upon numerical results of counting. Numerical data can be expressed in different scales that differ according to their levels of precision.

Nominal scales involve classification of each case into a distinct category. One form of nominal scale is the BINARY NOMINAL (dichotomous) in which there are only two categories. Examples would be ...

VARIABLE: Did the worker have an accident on the job?
BINARY NOMINAL MEASUREMENT: yes/no

VARIABLE: Was the customer a man or a woman?
BINARY NOMINAL MEASUREMENT: male/female

VARIABLE: Did the customer view the experimental ad?
BINARY NOMINAL MEASUREMENT: experimental/control

VARIABLE: Did the newly hired worker pass job training?
BINARY NOMINAL MEASUREMENT: passed/failed

Another form of nominal scale measurement is MULTIPLE NOMINAL, in which there are three or more categories. Here are some examples ...

VARIABLE: Which brand did the customer purchase?
MULTIPLE NOMINAL MEASUREMENT: brand X / brand Y / brand Z

VARIABLE: What type of job did the worker have?
MULTIPLE NOMINAL MEASUREMENT: clerical/production/sales

Hint: when dealing with spreadsheet programs (e.g., Minitab, Excel, SPSS) it is sometimes good to convert multiple nominal scales into a series of separate variable binary nominal scales. Religious affiliation (Catholic, Protestant, Jewish) would become three variables: Catholic yes/no, Protestant yes/no, Jewish yes/no.

ORDINAL scaling has some kind of ordering, seriation, ranking, or other comparison or gradation of magnitude or degree. In other words, ordinal scales mean that two cases can be compared in such a way so that one case can be said to have more of the variable, or be higher on that variable, compared to the other case.

VARIABLE: How do real estate agents rank on sales?
ORDINAL MEASUREMENT: first / second / third / ... / last

VARIABLE: Do workers agree with the new policy?
ORDINAL MEASUREMENT: agree / neutral / disagree

VARIABLE: Worker performance rating
ORDINAL MEASUREMENT: outstanding/good/fair/poor

VARIABLE: Years with company
ORDINAL MEASUREMENT: under 2 / 3-5 / 5-10 / over 10

VARIABLE: How often does customer have to wait in line?
ORDINAL MEASUREMENT: never / sometimes / usually / always

INTERVAL scaling uses numbers in such a way that there is a constancy of units: e.g., the distance from 3 to 4 is equal to the distance from 9 to 10.

VARIABLE: Temperature on a thermometer
INTERVAL MEASUREMENT: 75 degrees Fahrenheit

VARIABLE: Worker performance evaluation
INTERVAL MEASUREMENT: rating a 7 on a 1-10 scale

RATIO scaling requires everything that interval scaling has and also the requirements that the scale have a true zero point, and proportionality (e.g., 40 kilos is half the weight of 80 kilos).

VARIABLE: Worker production
RATIO MEASUREMENT: 321 units produced

VARIABLE: Sales of a branch office
RATIO MEASUREMENT: $21,635 last week

VARIABLE: Defective units discovered
RATIO MEASUREMENT: 3 units last shift

VARIABLE: Time
RATIO MEASUREMENT: the operation took 13.6 seconds

VARIABLE: Distance
RATIO MEASUREMENT: traveled 136 miles

One important distinction for interval and ratio scales is that some of them are discrete and others are continuous. DISCRETE variables have indivisible units, such as workers, units produced, accidents, trials passed. CONTINUOUS variables are ratio or interval scales with divisible units of things such as time, money, length, volume, area.

Hint: you could rank the scores, and have an ordinal scale, or put the scores into categories and have a nominal scale, but you would lose the precision of interval or ratio scales.

MISTAKE 7: Using a multiple nominal scale with many categories that have only a few subjects within them.

EXAMPLE: The variable of religious affiliation is multiple nominal. In the general U.S. population, no one denomination has more than about a third of the total, and there are many denominations that might only get one (or no) subjects in a particular sampling.

SOLUTION: One solution is to group several related categories: perhaps the Baptists with the Church of Christ, the Pentecostals with the Assembly of God. Of course, the more distinct denominations which you lump together under "Protestant" and "Christian," the more you are mixing together apples and oranges (as fruits).

Another solution is to draw the sample specifically from the categories which you wish to study. If you really want to compare Mormons and Jehovah's witnesses, do not assume that you can stand outside of a college library and from the sampling of students who come by, that you will get that many Jehovah's witnesses. A better approach is to stand outside of a Kingdom Hall and get the Jehovah's witness subsample, then stand outside of a Mormon ward and get that subsample.

MISTAKE 8: Asking for a binary nominal answer when the exact cut off point is unclear.

EXAMPLE: The frequency of exercise is on an ordinal scale. The following question attempts to reduce that to a binary nominal with a vague cut off point.

Do you exercise regularly?   YES   NO

SOLUTION: The cut off point can be stated with more precision.

During an average week, do you exercise vigorously (apart from the work you do) for at least a total of 60 minutes?

YES   NO

An alternative would be to employ an ordinal scale.

How often do you engage in vigorous physical exercise (apart from the work you do)?

NEVER

RARELY

ABOUT ONCE OR TWICE A WEEK

ALMOST DAILY

Chapter 3: Reliability and validity: how much to worry

Consistency of measurement

RELIABILITY means consistency of measurement. This is especially important in standardized psychological tests, but reliability is a criterion for any operational measure of a variable. Imagine that a twelve inch ruler were made out of elastic instead of wood or plastic. One carpenter might measure a board as being 5 inches, but another carpenter using the same ruler might stretch it a little less and determine that the board was 6 inches. This kind of inconsistency is not tolerable in science.

Establishing the reliability of a test	
Type of reliability	**Research involved**
Test-retest	Give the test twice to each subject; Correlate first administration to the second
Inter-rater	Have two judges evaluate each subject; Correlate the first ratings to the second
Alternate form	Give two versions of the test to each subject; Correlate the first version to the second
Internal	Give the entire test to each subject; Correlate one part of the test to the rest of it

One form of reliability is inter-rater. Two different raters (judges, interviewers, diagnosticians) evaluate the same subjects on the same variables. The more subjects on which the raters agree, the higher the inter-rater reliability. In the example below, the raters agree on 18 out of 20 cases, 90% of the time (compared to 50% which would be expected via random guessing).

Inter-rater Reliability

		SECOND RATER scores subject		
		High	Low	totals
F I R S T R A T E R	Subject is scored high on the variable	A 8 **AGREEMENT**	B 2 **DISAGREEMENT**	10
	Subject is scored low on the variable	C 0 **DISAGREEMENT**	D 10 **AGREEMENT**	10
	Totals	8	12	N = 20

Another form of reliability is test-retest. The subject is asked the same question on two different occasions to see if he answers consistently. Suppose we ask workers how important wages are, and we get inconsistent results: one week a worker rates wages as extremely important, and the next week that same worker rates wages as only moderately important. If many workers are varying their own responses on their question over time, the answers are not reliable.

Test-retest Reliability

		SECOND TIME: wages seen as		
		Very important	Less important	totals
F I R S T T T I M E	Very important	A 30 **AGREEMENT**	B 0 **DISAGREEMENT**	30
	Less important	C 0 **DISAGREEMENT**	D 20 **AGREEMENT**	10
	Totals	30	20	N=50

Other forms of reliability include alternate (parallel) forms in which there might be two slightly different versions of the same questionnaire, and internal reliability in which we applies especially to psychological tests in which many separate questions are added together to form a complete scale. (The different parts of the scale should really be measuring the same thing as the other parts of the scale.)

> **MISTAKE 9: Assuming that a measure is reliable instead of performing research to establish reliability.**

EXAMPLE: A human resources director was looking for the best predictors of future job performance of workers in the shipping and receiving department. She decided to ask the new department supervisor to rate the overall performance of the two dozen workers in the department. There was no guarantee that the supervisor's ratings would be consistent with any other measures of the workers' performance.

SOLUTION: Whenever we measure a variable by having a rater come up with a subjective, overall assessment of how the subjects rate on a particular variable, it is a good approach to attempt to establish inter-rater reliability. Suppose we could also get a rating of the same workers from the former department supervisor (who has just recently retired or been promoted). We could than compare the scores given by the two raters. If there was widespread disagreement (e.g., a correlation that was weak or even inverse) then this would cast doubt upon the reliability (and even the validity) of the operational definition. If the raters only disagreed on a few cases (say, four) but agreed on twenty cases (perhaps that ten were clearly good, and ten were clearly poor performers), we could then exclude those cases of disagreement from our final sample.

Note: Reliability does not imply consistency between the different subjects. We expect that different workers will have different scores on a given variable. Some workers may find wages to be extremely important, while other workers may rate wages as less important. We expect subjects to vary (and that is why we call these measurements variables). Reliability merely means that there should be some consistency in the answers that a given subject gives on a given variable.

VALIDITY means that a measurement actually measures the variable that it claims to measure. Validity and reliability are both important for scientific measures, but they are not the same thing. Imagine that you need to weigh a brick, and someone brings out a ruler. That ruler may measure very reliably (consistently) but what it measures is distance, not what we need to measure now, which is weight. One of the biggest problems in survey research is using the wrong tests to measure variables, and asking the wrong questions for the answers we are really looking for.

Going after the wrong variable

> **MISTAKE 10: *Asking a question for the subject to answer when a more valid answer can be obtained in a less confusing way.***

EXAMPLE: On a national telephone survey, is it really necessary to ask what state a person lives in when you just called the 909 area code of southern California?

EXAMPLE: I was supervising a team of student researchers in a nursing home. The questionnaire had to be administered orally, and the answer sheet was filled out by the researcher, not the subject. The first item dealt with the patient's sex. I had intended that the researcher would just look at the patient and circle "male" or "female" without even asking the patient. One young woman, however, went through each question on the list, giving the patient an opportunity to come up with an oral response. To her surprise and embarrassment, one 91 year old man responded to the question "Sex?" by saying "rarely, very rarely now."

SOLUTION: For face-to-face questionnaires, gender can be easily recorded by the questioner. Over the phone there might be a little more difficulty identifying the responded by gender (but there would be more embarrassment by asking a direct question, "Are you a man or a woman?"). A better approach might be to ask something like "This questionnaire is for women over the age of 18, is there anyone in your household who would fit that?" The subject would assume (and perhaps be flattered) that the examiner was questioning her age, not her gender.

> **MISTAKE 11: *Measuring a variable that appears related to the variable you really want to measure, when you could directly measure that variable.***

EXAMPLE: You have the task of figuring out which customers are most likely to close their accounts with Bank of America, so you ask.

How happy are you with the overall service provided by your bank, savings & loan or credit union?

VERY HAPPY   NOT TOO HAPPY   UNHAPPY

Should we infer that people who are unhappy are the ones who will close their accounts? Many Bank of America customers may answer "unhappy" but may remain with that institution simply because it is the only one that has a branch close.

EXAMPLE: You have the task of determining if the demand for Lincoln automobiles will increase.

Which of these luxury automobiles would you prefer?

LINCOLN   CADILLAC   CHRYSLER   LEXUS   MERCEDES

I would have no trouble answering that question. I love my 1980 Lincoln Mark VI, and spend thousands of dollars a year keeping it running well and looking good. Does that mean that I am in the market for a new Lincoln? No, I prefer the classic looks of my old Lincoln.

Someone else might circle Lincoln as the choice, but that may be more of a dream car, and she may end up settling for a Ford Crown Victoria.

SOLUTION: Directly measure the variable that you want to measure. Usually that means assessing the probability of some future action on the part of the subject (worker or consumer). This estimate is improved when a time frame is given.

How likely is it that you will close your Bank of America account within the next twelve months?

VERY LIKELY   SOMEWHAT LIKELY   NOT VERY LIKELY

How likely is it that you will purchase a new Lincoln within the next twelve months?

VERY LIKELY   SOMEWHAT LIKELY   NOT VERY LIKELY

MISTAKE 12: *The use of scales that evaluate but do not suggest improvement.*

EXAMPLE: An automobile dealer gave a written questionnaire to those who had taken a test drive in a small truck.

24

What do you think about the trade off between fuel economy and performance?

The truck was rated very low. What would it take to get the customers to purchase the truck? better mileage? a bigger engine with quicker acceleration?

Faculty evaluations at the small college where I teach give a list of favorable characteristics of the professor, and the student has to fill out a bubble form indicating how well that professor has that characteristic. One of the items referred to the wise management of class time. One particular adjunct faculty member scored low on this item. We knew there was a problem, but the solution was not evident. Was she supposed to lecture more or have more discussion? Another question dealt with whether or not the level at which the course was taught was appropriate. Students graded her poorly on that item as well. Does that mean she needs to come up a level or go down a level?

SOLUTION: Balance scales are appropriate when there is some golden mean that should be sought, and it is wise to avoid both extremes. These are sometimes known as "Goldilocks" scales, because we are seeking to avoid that which is too much on one extreme, too much on the other, and get that which is just right. The automobile dealer could have used a question such as

All vehicles involve a trade-off in terms of fuel economy and performance (e.g., acceleration). How would you rate the XYZ model you just drove?

| NEEDS BETTER | JUST ABOUT | NEEDS BETTER |
| FUEL ECONOMY | RIGHT | ACCELERATION |

That information would be useful in redesigning the truck (if the overwhelming answer is one side of the other). If both extremes get a large number of votes, and few are in the middle, there may be a need to develop two separate models: a high performance version (perhaps targeted at younger males), and a fuel efficient version.

Let's go back to the faculty evaluation questions, which could be rephrased as follows.

What is your impression of the balance between lecture and discussion in this course?

TOO MUCH LECTURE   JUST ABOUT   TOO MUCH DISCUSSION

TOO LITTLE DISCUSSION   THE RIGHT MIX   TOO LITTLE LECTURE

What do you think about the general level of this course?

TOO ADVANCED   JUST ABOUT   TOO MUCH

FOR ME   THE RIGHT LEVEL   LIKE HIGH SCHOOL

The answer to this latter question might indicate a need to redesign the course, making it more difficult or easy. If both extreme answers receive high percentages compared to the middle, there might be a need for separate courses: one for majors and one for non-majors.

Overlooking confounding variables

> **MISTAKE 13:** *Asking a subject for a self-evaluation of performance tends to result in over-inflated estimates.*

EXAMPLE: College professors were asked to assess their own individual teaching effectiveness.

How do you think your teaching ability rates compared to that of most professors in your field?

ABOVE AVERAGE   ABOUT AVERAGE   BELOW AVERAGE

Over three quarters said above average, while no one said below average.

SOLUTION: This is clearly a case where an objective measure (e.g., how well the professors' students do on a standardized test) might be a more valid measure of the variable. Having students, or deans, or even peer colleagues rate the performance of professors would also open up the possibility of inter-rater reliability.

> **MISTAKE 14:** *Many researchers who develop psychological tests or questions on a survey know that they are not directly measuring the variable that they really want to measure, but operate under the assumption that the data obtained will nevertheless allow them to make some inference about the real target variable.*

EXAMPLE: One human resource director did not want to take the expense of developing a valid and reliable test to assess the kind of mechanical aptitude necessary for learning how to perform well on an assembly line job. He decided to give each job applicant a short IQ test. This proved not to be a valid measure of mechanical ability, since most IQ tests measure mathematical and verbal skills, rather than mechanical aptitude. (Since IQ tests also lack cross cultural fairness, they may also be illegal if that results in adverse impact on the hiring of categories of applicants protected by the Equal Employment Opportunity Commission.)

SOLUTION: The validity of any predictor of future employment should be established by research that correlates the pre-employment predictor to post-employment performance. In general, the best predictors of mechanical ability are not paper and pencil tests, but hands-on demonstrations, and previous experience.

The error of aggregation

> *MISTAKE 15: Assuming that a list of questions with individual face validity can be aggregated into a scale that necessarily has validity and reliability.*

EXAMPLE: An investigation of managerial effectiveness had hypothesized that managers who had earned an M.B.A. degree would be more effective than those without such degrees. The researcher created ten statements describing what she believed to be the effective manager, and rewrote each of these as a self-description with Likert response format. Here are two examples.

> I am almost always open to my subordinate's suggestions.

> I make efficient use of the productive resources at my disposal.

The researcher simply gave a "strongly agree" three points, a "mostly agree" two points, a "mostly disagree" one point, and a "strongly disagree" zero points, then added everything up for a range of zero to forty. Even if some of these self-evaluations were valid measures of managerial effectiveness, were they all valid (and equally so, as the scoring assumes)?

SOLUTION: Before we attempt to correlate a variable (e.g., managerial effectiveness) to another variable (e.g., having an M.B.A. degree) we must start off with a valid and reliable measure of each of the variables concerned. A preliminary study should have validated the measure of managerial effectiveness, by taking each of the ten items correlated with some external, established standard of the variable (e.g., a senior executive rating of managerial effectiveness, objective measures of performance such as ROI).

Assuming validation before research is done

> *MISTAKE 16: Assuming that you can validate a new scale and use it in the same research. The validation of a scale requires at least one separate, preliminary stage of research.*

EXAMPLE: A trucking company needed to come up with a valid predictor of which job applicants would be mostly likely to become safe and responsible future drivers. One of the supervisors came up with a list of ten statements that he thought should identify good prospects. He then gave this test to a sample of two dozen applicants, and correlated these answers with background variables such as previous training and experience. He then assumed that the those background factors which correlated with the ten item scale would also be valid predictors of good drivers.

SOLUTION: A scale must be validated in one phase of research, and preferably re-validated in a second phase of research, before we attempt to use the scale in research on the variable it tries to measure. Ideally, the validation of the above scale should have taken two phases of research before it was employed in additional research.

The first phase could have been a cross-sectional survey. These ten questions could have been asked to present (and/or former) drivers. We would also need to have some other measure of the drivers' performance (e.g., archives with safety and attendance records, supervisory ratings).

Cross Sectional Validation			
(subjects are present drivers)			

		ESTABLISHED MEASURE		
		GOOD driver	BAD driver	totals
S C A L E	Driver answers YES	A **AGREEMENT**	B **DISAGREEMENT**	
I T E M	Driver answers NO	C **DISAGREEMENT**	D **AGREEMENT**	
	Totals			N

On those items where we have a pattern of the vast majority of subjects ending up in cells A and D, that is a valid test item. We can use it by selecting those drivers who answer YES to the pre-employment question. On those items where we have a pattern of the vast majority of subjects ending up in cells B and C, that is a valid test item. We can use it by selecting those drivers who answer NO to the pre-employment question. If the proportion of A / (A+B) is similar to the proportion of C / (C+D), then there is no useful pattern of responses.

Then, a second, revalidation stage of research would also take place. Take those items that passed the first, initial (cross sectional) stage of validation and give this scale to new applicants who are then hired. Then come back after a year or so and measure these future drivers' performance (with supervisory ratings or objective data on safety and attendance from the employment records). This type of research is known as longitudinal because it takes place over a long time period.

Longitudinal Validation
(subjects are future drivers)

		ESTABLISHED MEASURE		
		GOOD driver	BAD driver	totals
S C A L E I T E M	Applicant answers YES	A **AGREEMENT**	B **DISAGREEMENT**	
	Applicant answers NO	C **DISAGREEMENT**	D **AGREEMENT**	
	Totals			N

Then, ideally, a third stage of re-validation occurs. A new crop of applicants will take a scale composed of those items that have passed both the first and second validations. Each applicant receives a score based upon the number of items "passed" (so, if there were ten items on the scale, each applicant could get a score ranging from 0 to 10). Then we wait for another year or so and measure these future drivers' on the job performance, classifying each of them as a good or bad driver. We then look for the cut off score that offers the highest correlation with the outcome performance measures. In other words, we look for the cut off score (say 7 and above) that will result in the fewest false positives and false negatives.

A false positive is an applicant who scored well on the test (he looked good on paper), but his performance ended up being poor. A false negative is an applicant who scored poorly on the test, but ended up doing well where it counted, on the job. While both false positives and false negatives are to be avoided, there are times when one is worse than the other. When there are many job applicants compared to the number of available positions, false negatives are not a problem, and so we just need to find a test and scoring which will eliminate the false positives (i.e., assure that every one hired will perform well on the job). An organization concerned about affirmative action issues must be cautious that its selection techniques do not result in false negatives that demonstrate adverse impact against protected classes (e.g., women, older workers, disabled workers, ethnic groups).

Longitudinal Validation
(subjects are future drivers)

| | | ESTABLISHED MEASURE | | |
		GOOD driver	BAD driver	totals
S C A L E	Applicant scores above cutoff	A **TRUE POSITIVE**	B **FALSE POSITIVE**	
T O T A L	Applicant scores below cutoff	C **FALSE NEGATIVE**	D **TRUE NEGATIVE**	
	Totals			N

MISTAKE 17: Constructing a scale to be validated, when other predictors can be validated directly from archival data.

EXAMPLE: Consider again the above example of the trucking company.

Was it really necessary to go through these three steps of scale validation in order to have a valid predictor for future on the job performance?

SOLUTION: Just using the last step of longitudinal research would be sufficient to get valid predictors if those items came directly from background factors present at the time of hiring and recording in the employment records. Two of the best predictors would be previous training and previous job experience. Suppose instead of looking at some scale on a questionnaire, we looked at number of years of experience prior to joining this firm as a truck driver.

Longitudinal Validation
(subjects are future drivers)

| | | ESTABLISHED MEASURE | | |
		GOOD driver	BAD driver	totals
PRIOR EXPERIENCE	Applicant has more than cutoff	A **TRUE POSITIVE**	B **FALSE POSITIVE**	
	Applicant has less than cutoff	C **FALSE NEGATIVE**	D **TRUE NEGATIVE**	
	Totals			N

Rather than there being a linear relationship between number of years of previous experience and on the job performance, it is best to look for a cutoff point, say three years. Of course, this could also be done more quickly using cross sectional data from existing drivers, but it may not be as valid.

Whether or not the use of a questionnaire for job applicants would be more valid (i.e., have fewer false negatives and false positives) than the use of these background factors is an empirical question. Ideally, the trucking company should develop many valid predictors of future performance.

The temptation of the shortened scale

> **MISTAKE 18:** *Assuming that a shortened, or otherwise modified scale will have the same validity and reliability as the original scale.*
>
> It is very tempting to say "This one item will not fit my population" or "This scale is too long, so we will just use half of the items." All of the previous studies establishing the validity and reliability of the scale used the original, longer form. More importantly, giving a sample a modified scale means that the change of comparing the sample to pre-established norms has been forfeited.

EXAMPLE: Several items on a test of fire fighting knowledge were deleted from a questionnaire given to recruits at a state department of forestry because those items did not relate to the types of fires encountered on public lands as opposed to the building fires encountered by most city departments. This is alright as long as there are no assumptions made about the internal validity of the modified test, or attempts to compare the overall scores of the forestry recruits to local city fire departments. Of course forestry scores will be lower: they had fewer items. If the purpose of the research was to show that forestry recruits had lower knowledge in certain areas, then it is necessary to include those areas on both the questionnaires given to the forestry and the city groups.

EXAMPLE: When I lived in Guadalajara, I decided to modify the Mental Status Questionnaire (a dementia screening test) to meet the context of Mexican elders. I also decided to come up with two versions: one for elders still living in the general community, and those who had be placed in nursing homes. Once I had made these modifications, I could no longer compare Mexican elders to the New York sample on which the original MSQ had been standardized, nor could I compare community aged to institutionalized aged.

SOLUTION: If a test is to be modified in order to make it more relevant to a new population, the old validity and reliability norms no longer apply. We are now obligated to perform new research in order to validate the new scale.

In the case of the modified MSQ (or International Version of the Mental Status Questionnaire) we validated it item by item by comparing the subjects' answers to the nursing home staff's assessment of the patient as "lucid and alert" or "consistently confused." The items that we included in the final version of the IVMSQ were those in which the lucid and alert patients got the right answer, and the confused patients got the wrong answer. We then used another validation technique on both the community and institutional version of the IVMSQ: correlating scores with other accepted measures of confusion (such as neurological tests).

Careless responses

> **MISTAKE 19: *Not catching careless responses.***
>
> This becomes a major problem because their data will greatly distort correlations and significance. While haphazard responding patterns tend to lower the levels of correlation and hurt statistical significance (leading to Type II error), subjects who show a consistent pattern of responses (e.g., always selecting YES answers or HIGH end answers) tend to strengthen correlations and give us better statistical significance (leading to Type I error).

EXAMPLE: Most of us have had the following experience on the other side of the questionnaire. We agree to fill out a brief questionnaire, and then notice that it is longer than we expected. If no one is around, we just throw it away. If someone is around, we are too embarrassed to admit that we will not fulfill the obligation of completing the questionnaire, so we just start circling answers, checking boxes, and then hand it in.

SOLUTION: The first and best solution is prevention: make the questionnaire so short and simple that it can be filled out quickly. Unintentional errors can be reduced by some of the response formats discussed in chapter 8. There are also some tricks that can be used to identify careless responders, so that their questionnaires can be removed, and their non valid data will not contaminate the results.

One approach is to include questions which will allow us to ascertain internal reliability (which is essential, though not adequate for validity).

At what age did you graduate from high school?

NOT YET   18 OR YOUNGER   19   20s   30s   40s   50s

At what age did you graduate from college with a bachelor's degree?

NOT YET   18 OR YOUNGER   19   20s   30s   40s   50s

At what age did you get your (first) graduate or professional degree?

NOT YET   18 OR YOUNGER   19   20s   30s   40s   50s

If someone claims to have a master's degree, but no bachelors, that is an extremely unlikely situation. Similarly, if someone finished high school at a later age than college, that is extremely unlikely. We would be best to assume in such cases that the responses given are not valid, but are due to carelessness.

Another example would be to ask questions about marital status.

Have you ever been married?

YES   NO

Have you ever been divorced?

YES   NO

It is possible for a subject to answer NO to both questions, YES to both questions, or (in my case) YES to the first and NO to the second. However, it is not possible to answer NO to the first and YES to the second. Such a response would represent non valid data and should be removed: we should infer that the responses on other variables also have suspect validity.

The teaching evaluations at Wichita State University are given every term to every class. Those tabulating the data rightly worry that students might get tired of the process and just start filling in the bubbles on the answer sheets. In order to catch such non valid responses, the questionnaire has built into it several items for catching careless responders. One item tells the subject to leave that row of answers completely blank. Another item tells the subject to fill in any TWO answers on that row (instead of the customary one). The machine which tabulates the forms is programmed to look at these two validity items first, and throw out any form which has either of these items marked incorrectly.

Chapter 4: How to prove it: the four basic designs

In order to understand how surveys and experiments test hypotheses, we must clarify these terms: group, sample, and population.

Part / Whole Relationships			
P O P U L A T I O N	S A M P L E	First group (women)	Sally Jones (one subject)
			Maria Garcia (one subject)
			Betty Williams (one subject)
			Michelle Nguyen (one subject)
		Second group (men)	Bob Smith (one subject)
			Juan Gonzalez (one subject)
			Bill Johnson (one subject)
			Eric Wong (one subject)
	Subjects not observed		

Subjects are the people actually observed and from whom data are gathered about the variables in question. Subjects are people whose dependent variables (e.g., performance, attitudes) are being measured. In business related research, the subjects are usually workers (or job applicants) or customers (or potential customers).

Population refers to the type of subjects being studied, such as all workers in a given occupation, or all users of a certain product.

Sample refers to those specific subjects actually observed in the research, such as the workers who participated in special job training, or those customers who actually filled out a questionnaire on the back of the warranty registration card.

When we speak of a specific group we mean a part of the sample that differs from the rest of the sample, such as those workers who were males, or those customers who received an advertisement for a rebate.

> *MISTAKE 20: Using the terms "population," "sample," and "group" interchangeably as if they were synonymous.*

EXAMPLE: One student reported "We took a population of ten rats and put them in one of two samples: experimental or control."

SOLUTION: Remember that "population" refers to the entire class of subjects studied: in this case, all rats of a given species. Remember that "sample" refers to all the subjects actually observed in a particular research project: in this case, all ten rats. Experiments then randomly assign each subject into one of several "groups" which are treated differently. Surveys merely measure pre-existing "groups."

Whenever we are trying to "prove" something, we must select one of four different research designs.

Design	Basic concept	Advantage	Limits
One sample	Compare entire sample to some external norms	Quick, requires smaller sample size	Norms must exist, and we assume that our sample does not differ from the norms on any important background variable (except for the one we are studying)
Correlational	Measure two variables and look for a relationship	Quick, requires no external norms	The causal relationship can be easily misinterpreted
Separate groups	Divide sample into two comparison groups	Requires no external norms	For surveys, same as correlational; for experiments, the grouping must be by random assignment
Repeated measures	Measure each subject twice on the same variable	Controls for inter-subject variability; requires smaller sample size	May take a long time; need to match data; effects of attrition, practice, fatigue, etc.

Sample vs. Norms

One sample designs take the entire sample and compare it to some pre-established norms (such as means or percents for the entire population).

EXAMPLE: Are warehouse workers more likely to have an industrial accident? Compare a sample of warehouse workers to the company-wide accident rate.

> ### MISTAKE 21: Assuming that the norms are 50/50 in a one sample design.
>
> One sample designs are also known as sample vs. norms because they compare the entire sample's central tendency (i.e., percent or mean) to that of some external norm (e.g., the central tendency of a relevant population. If those population norms do not exist, then we cannot conclude that a sample has scored high or low because we have nothing to compare it with.

A 50/50 norm may be appropriate for some variables. We can assume that the U.S. population is roughly half male and half female, and therefore an observed sample that is mostly male differs from those norms. We can assume that a flipped coin should have a 50/50 chance of coming up heads. If we observed that it comes up heads ten times out of ten, it is improbable that it represents an honest coin flip.

EXAMPLE: A gambler falsely assumed that his odds at winning on a throw of the dice were 50/50. (The odds are actually weighted slightly against the thrower.)

EXAMPLE: A nursing home did a field count of the patients who came morning exercises. Fifteen patients showed up, and all but two were women. Assuming a 50/50 norm, this looked like a significant difference.

EXAMPLE: One company did a survey of the job satisfaction of its employees. The results indicated that 60% were satisfied. The sample size was large enough to that these data were significantly higher than a norm of 50%.

SOLUTION: The exact norms must be used. Usually these can be found in census records or in national polls.

In the above case of the gambler, he should go into any game of pure chance with a realistic understanding of the odds against him. Unfortunately, too many gamblers assume that the laws of probability guarantee a winning streak after a losing streak. All that the laws of probability guarantee is that over the long run, the more that is gambled, the more that will be lost.

In the case of the nursing home, the relevant population here is not that of the entire U.S., but those people actually in the nursing home. In this particular facility, it was three quarters women. The male/female breakdown of the sample that showed up for the morning exercises was very proportionate to the population of the nursing home.

The company that measured 60% job satisfaction could have consulted some national polls, and found that they consistently show that 80% to 90% of Americans express satisfaction with their jobs (because those who are not satisfied, end up looking for more satisfying work, and keep on looking until they find something more tolerable). So, a company with 60% job satisfaction could be significantly below the national norms.

MISTAKE 22: The use of dated norms.

Outside of political polls, most polls are several months old by the time they get published. It might be several years before a national polling organization gets around to a particular topic again. U.S. Census figures may not be updated for ten years. The actual population norms may have changed between the time the population was last measured and the time when your sample was surveyed.

EXAMPLE: A local ISP provider wanted to see if internet use was particularly prevalent in his city. He found a national survey showing that 40% of U.S. households were connected to the internet. He was overjoyed when his sample showed a 60% rate. The problem was that the national survey came from three years ago, and the 60% reported by his sample may be very close to the current national norm.

SOLUTION: If you are doing research on a topic that has been changing over the last few years, get the most recent data on national norms. If you suspect that your norms are dated, then choose some other design to test your hypothesis: separate groups, repeated measures or correlational.

> **MISTAKE 23: Using national norms that are inappropriate for a local population.**

EXAMPLE: Does Starbuck's attract a disproportionately female clientele? A survey of one Starbuck's in Manhattan noticed that over two-thirds of lunch time clientele was female. The situation was that the office workers of downtown Manhattan are not a 50/50 split of males and females, but are disproportionately female.

EXAMPLE: Does Starbuck's attract a disproportionately older clientele? A survey of one Starbuck's in Sun City, California noticed that the majority of the clients were over age 65, while only 12% of Americans are in this age category. The situation was that Sun City is largely a retirement community, and the vicinity around this Starbucks is mostly housing which requires a minimum age of 55 to move in.

EXAMPLE: Does Starbuck's attract a disproportionately Asian clientele? A survey of one Starbuck's in Riverside, California noticed that the majority of the clients were Asian, while only 12% of Californians are of Asian ancestry. The situation was this Starbuck's was located next to a University of California campus on which a plurality of the students are of Asian ancestry.

SOLUTION: If you are trying to show that your local geographical area differs significantly from the national norms, then the sample vs. norms design will work for you. If you are trying to test some other hypothesis, then geography becomes a powerful confounding variable because most census data and polling data come from across the nation while your sample of workers or customers tends to be local. This is another reason for considering some other design to test your hypotheses: separate groups, repeated measures or correlational.

Correlational

Correlational designs measure two variables and then determine if there is any association between them.

EXAMPLE: Are older warehouse workers more likely to have industrial accidents? Measure each worker's age and number of accidents and look for a trend.

Correlational designs are pretty simple and straight forward. As long as there are no problems in the measurement of each variable (discussed in Chapter 2) or in interpretation (discussed in Chapter 11) the results are useful.

Separate groups

Separate groups designs take the sample and divide it up into two or more groups, and then compare the groups.

EXAMPLE: Are warehouse workers more likely to have an industrial accident than assembly line workers? Get a sample that includes both types (groups) of workers and compare the accident rates of each group.

Especially when there are just two groups (e.g., men and women, experimental and control), the design is fairly simple. The greatest vulnerability of this design is an initial grouping that admits confounding variables (to be discussed in Chapter 5).

Repeated Measures

Repeated measures designs measure the sample on the same variable at two different points in time (or on two different aspects of the same variable) and see if there is a significant increase or decrease. One of the great advantages of repeated measures designs (over separate groups or correlational) is that we end up controlling for inter-subject variation and that means that we can get better statistical significance with a smaller sample size).

EXAMPLE: Did warehouse workers significantly reduce accidents after training? Get a sample of warehouse workers, put them all through training, and measure their accident rate before and after training.

Another type of repeated measures design is when we have two raters of the same variable per subject.

EXAMPLE: Suppose we have a hypothesis that men are more satisfied with their marriages than women are. Our sample might be composed of two dozen married couples. Within each couple, we measure the husband's reported level of satisfaction, and the wife's reported level of satisfaction. We compare within each married pair, and look for a trend of the wives' ratings being higher than that of their respective husbands.

> ## MISTAKE 24: Retrospective and subjective measures of a variable.
>
> The best way to measure these two time periods is prospectively (longitudinally). We measure the variable now, and then come back later and measure the variable again. A more convenient (but less valid) approach to measurement is a retrospective design: we look at present data and then go looking for measures of the past situation. This is especially problematic when we ask the

subjects to estimate their own previous level. The data for the past may be distorted by poor memory or by the subject's own motivation.

EXAMPLE:

How effective are you in your job, now that you have been trained?

VERY   SOMEWHAT   NOT VERY   NOT AT ALL

How effective were you in your job, prior to being trained?

VERY   SOMEWHAT   NOT VERY   NOT AT ALL

SOLUTION: When we have to use retrospective data, objective measures (e.g., production or accident records) are superior to subjective estimates.

MISTAKE 25: Using a longitudinal repeated measures design, but changing the way that you measure the variable between the first measure and the second measure. This results in an "apples and oranges" comparison.

EXAMPLE: A medical research team wanted to see if patients got more depressed after spending a year in a nursing home. Twenty newly admitted patients were assessed for depression using the Beck Depression Inventory. A year later, there was a follow up measure of depression, but it was decided to use the Geriatric Depression Scale because it was more valid and reliable than the Beck. The decision to switch depression scales probably resulted in a better follow up assessment of depression, but it also precluded a comparison with the previous measure. There is no way to convert GDS scores to BDI the same way that Celsius can be converted to Fahrenheit.

SOLUTION: If you are doing a repeated measures design, and you want to have a follow-up measure which can be compared with the initial measure, you must stick with the initial measure, even though certain improvements could be made.

> **MISTAKE 26: Using a repeated measures design, but ignoring some changing factors which will influence the dependent variable at follow up.**

EXAMPLE: A company wanted to see if training would influence productivity. Each worker's February 2000 output was measured, and then in June each worker underwent three months of training for thirty minutes a day. Each worker's February 2001 output was measured again, but to the trainer's dismay, output had declined. Fortunately, the CEO did not blame the training, but noted that the company had a marked decrease in orders, and there was simply less work to be done.

SOLUTION: Always be aware of the impact of such external factors that influence the dependent variables. These factors can never been completely controlled or assessed.

> **MISTAKE 27: Sampling from the same population twice and assuming that you have a repeated measures design.**
>
> The key aspect of a repeated measures design is that the same variables are measured again in the same sample, not two similar samples drawn from the same population.

EXAMPLE: One real estate firm measured the annual income of its customers in 1999 and again in 2001 and noticed that the median income was 50% higher in 2001. This does not mean that the firm's 1999 customers made a lot more money when they were measured again two years later. The 2001 customers were a different group of people, a different group with much higher incomes.

EXAMPLE: Of ten adult customers coming out of a grocery store at 11 AM, eight were women. Of ten adult customers coming out of the same store at 6 PM, only four were women. The fact that it was the same store does not make it repeated measures. Even if some of the customers noticed at 11 AM were also noticed at 6 PM, it is not repeated measures unless the entire sample is measured twice.

EXAMPLE: One company measured the job satisfaction of its workers in 1999 and again in 2001 and noticed an increase. This does not mean that the same workers who were dissatisfied in 1999 had changed their minds by 2001. There may be a large overlap between the two groups (most workers may have been present for both the 1999 and the 2001 surveys), but some of the 1999 workers did not give data in 2001 because they died, quit, retired, were laid off or fired. Some of the 2001 workers were recent hires, and so no 1999 data exist for them. Whenever there is rapid turnover (which tends to be associated with poor satisfaction) or rapid expansion the overlap between the two groups will be lower.

SOLUTION: The above research examples give valuable information and can test hypotheses, but not with a repeated measures design. Regard each of the above examples as a separate groups design. The real estate sample has two groups of customers: those who dealt with the firm in 1999 and those who dealt with the firm in 2001. The grocery store sample has two groups: those who came through in the morning and those who came through in the late afternoon. The sample of workers has two groups: those working in 1999 and those working in 2001.

> *MISTAKE 28: Performing a repeated measures design, but not keeping track of which subject's prior measure is associated with the subsequent measure.*
>
> The end result is that you have two piles of questionnaires, one from before and one from after. The piles may be equal in number and may represent the exact same subjects, but if you do not know which questionnaire in the first pile is to be matched with which questionnaire in the second pile, you cannot do a repeated measures design.

This problem becomes most apparent when you try to put your data into the computer programs designed to handle repeated measures data. You will be asked for a given subject's prior data, and then that same subject's subsequent data. If you cannot enter that data in this paired format, the program cannot calculate statistical significance.

EXAMPLE: One type of repeated measures design is where we get two measures (one from each person in a matched pair) on the same variable. Two dozen married couples were asked to rate the quality of their marriage. The hypothesis was that most husbands would rate the marriage better than their wives had rated the marriage. The questionnaire had a place to indicate if the subject was male or female (and so a questionnaire could be identified as that of a husband or wife) but there was no way to match each husband's questionnaire with that of his particular wife. It would be inappropriate to pair them at random, because that would give a direct comparison of Mr. Smith's perspective on his marriage with Mrs. Jones' perspective on her marriage: another case of apples and oranges.

EXAMPLE: One company measured the job satisfaction of its workers in 1999 and again in 2001. During that time, the work force remained the same. No one was hired, fired, quit, retired, laid off or died. However, because the surveys were anonymous, there was no way to determine which worker had filled out the 1999 form to be matched with the 2001 form. So it was impossible to state what percent of workers had increased their job satisfaction, and it was impossible to use a repeated measures design to test for significance.

SOLUTION: The above examples still provide useful data, but the data can only test hypotheses with a separate groups design.

The marriage study cited above can tell us if the mean level of satisfaction of the group of husbands is significantly higher than the mean level of satisfaction of the group of wives. We cannot answer the question if the majority of husbands are more satisfied than their wives because we cannot identify individual couple pairings.

The job satisfaction study cited above can tell us if the mean level of satisfaction of the 2001 workers is significantly higher than that of the 1999 workers, but we cannot determine what percentage of workers increased their satisfaction.

The only way to assure that we can analyze the data in a repeated measures design is to track all measures by the same subject (or pair). This tracking should be done in such a way as to minimally compromise anonymity. For example, in the case of the paired questionnaires going to husband and wife, make a similar mark on the back of each of the two questionnaires, thus identifying the Smiths as couple A and the Jones and couple B, etc. An even less obtrusive technique would be to mark each pair of questionnaires on the back with a vertical line of the same height. Suppose you had two dozen couples. One couple would get a pair of questionnaires marked with a line on the back one centimeter from the bottom. The next couple would get a pair of questionnaires with a small mark two centimeters up from the bottom. The important thing is that when we get back the questionnaires we can tell which questionnaire came from the husband, which came from the wife, and that this husband and wife pair go together.

MISTAKE 29: Attempting to include partial data in a repeated measures design.

EXAMPLE: A study of marital satisfaction used coded pairs of questionnaires distributed to twenty couples. Among the returned forms, 18 complete couples could be identified. One husband's form was found without that of his wife. One wife's form was found without that of her husband. So, we have some data for each of the twenty couples, but complete data on only 18. (We cannot put together the extra husband with the extra wife, because they are rating different marriages.)

SOLUTION: We must eliminate those subjects (or pairs) on whom the data are incomplete. So, our final sample size would only be 18.

If the problem of incomplete data is widespread, and the sample size would be reduced substantially by eliminating all incomplete data, an alternative would be go back to the less powerful separate groups design: just comparing all the husbands to all of the wives.

Indeed, both designs could be used. Suppose we had only 19 husbands respond, and only 13 wives.

"The 19 men reported a mean level of satisfaction of 8.3 while the 13 wives who responded reported a mean level of satisfaction of 7.2 (p < .05). Twelve complete husband-wife couple pairs could be identified from the data, and in all but one of these, the husband reported greater satisfaction (p < .05)."

> ## MISTAKE 30: *Ignoring the impact of present time frame on both retrospective and prospective repeated measures designs.*
>
> A retrospective survey asks subjects to remember what it was like in the past. (This is supposed to take the place of actually having previously measured the variable.) A prospective study asks subjects to imagine what it will be like at some point in the future. (This is supposed to take the place of having to subsequently measure the variable.)

One of the easiest ways around the problem of tracking repeated measures responses is to put both of the responses on one piece of paper filled out by the same person at the same point in time. Unfortunately, an individual's own perspective at the present colors what that person imagines his own future (or past, or spouse's) perspective will be (or was, or is).

EXAMPLE: A car dealer was trying to see if car owners had a rapidly decreasing level of satisfaction.

"How satisfied are you with your car today?"

"How satisfied were you with your car when you purchased it?"

"How satisfied do you think you will be with your car in another year?"

We must seriously question the validity of this process. The validity of both past memories and future predictions are easily confounding by present anticipations.

> ## MISTAKE 31: *Assuming that having one member of a pair can give an estimate of the rating of the other member of the pair.*

EXAMPLE: In order to see if men are generally more satisfied with their marriages than their wives are, twenty men were asked the following questions.

"How would you rate your satisfaction with your marriage?"

"How do you think that your wife would rate her satisfaction with your marriage?"

SOLUTION: The problem with the above items is to assume that they are valid measures of the real past, future, or spousal perspective. If these questions are to be used, let us recognize that they are just perspectives, and not objective measures. It would be most helpful to ask the husband to predict his wife's satisfaction, and then compare that with the actual level of satisfaction reported by the wife. This gets us back into the problem of keeping track of the couple's responses, but it would help us test the hypothesis that most men over-predict their wives' marital satisfaction.

> **MISTAKE 32:** *Ignoring the impact of attrition on repeated measures designs.*
>
> The longer the time interval between the initial data gathering and the subsequent data gathering, the fewer members of the original sample can be surveyed again. If this attrition is differential, and impacts one end of the dependent variable more than the other, this can exaggerate, distort, or hide long term trends.

EXAMPLE: Worker satisfaction was measured in 1999 and again in 2001. Each questionnaire was carefully coded so that the 1999 data could be matched to the 2001 data. Of the hundred workers surveyed in 1999, ten quit, five retired, three were fired, and two died. The majority of the 80 subjects who were followed up showed a slight increase in job satisfaction, but the data were not statistically significant.

Attrition may have played a role in this increased satisfaction. The twenty workers who were not around for retest probably had lower initial satisfaction and may have been on a downward trend.

EXAMPLE: Attitudes about marriage were measured in fifty married men in their twenties and then forty years later. At follow up, ten men had died, five could not be found, and twenty-five had divorced their original wives, yielding a final sample of only ten who were still married to the same woman. All then of them reported more commitment to their marriages in their sixties than they had had in their twenties.

Should we conclude that men make better husbands as they age? The men who became worse husbands got divorced or disappeared.

SOLUTION: When there is high attrition in the original sample, consider supplementing the longitudinal design with a cross-sectional design in which the aging group is compared to a younger group tested at the same time.

> MISTAKE 33: *Ignoring the impact of practice effect on repeated measures designs. This is a serious problem when the dependent variable is a measure of performance and we have some reason to believe that the previous testing will provide practice that improves that performance.*

EXAMPLE: Suppose we want to prove that workers will perform better on an assembly task using special gripping gloves rather than bare handed. We present the workers with the assembly task and measure how fast they do it bare handed. Then we have them try it again wearing the new gloves, and we see that most of the workers demonstrate a substantial improvement in their performance.

It is possible that the gloves really do help, but a confounding variable is that the second time (with the gloves) the workers were performing the task, they benefited from the practice they had on the previous trial (without the gloves).

SOLUTION: One solution is to establish a stable baseline in the control condition before the experimental condition is employed. In the example of the gloves, have the workers keep on working in a bare handed condition, until they have sufficiently practiced with the task that their performance has stabilized and does not show additional improvement due to practice each time the task is repeated.

Another solution is to use a counterbalanced presentation. Half of the subject start with the bare hand, and then use the gloves; while the other half of the subjects start with the gloves, and then go bare handed.

Another solution is simply to have two separate groups of workers: those assigned to the bare handed condition and those assigned to the gloved condition.

> *MISTAKE 34: Ignoring the novelty effect of the experimental condition in a repeated measures design. When subjects first try something new, they might be overly impressed and express an extremely favorable attitude toward it, which will fade as the novelty wears off. On performance measures, novelty can lead to either temporarily increased motivation, or greater task difficulty.*

EXAMPLE: A bartender encouraged his customers to try a new citrus flavored alcoholic beverage. Sales jumped the first month, and then steadily declined. A year later, he noticed that no customers still had that drink as their regular.

EXAMPLE: In 1995 an office manager decided to switch word processing programs from one that was DOS based to one based in Windows-95. At first, there was a decrease in the quantity and quality of the documents produced. Fortunately, the office manager did not immediately decide to reverse course, but gave the secretaries time to get used to the new software. Within two months, quantity and quality was better than the pre-Windows level.

SOLUTION: When novelty effect is a possibility, give the experimental condition sufficient time for the novelty to wear off and the actual level of interest or performance can be evaluated.

> **MISTAKE 35:** *Ignoring the impact of natural trends within the variable being measured. Most repeated measures designs assume stability of the dependent variable over time, and therefore, any change in the dependent variable must be attributed to the impact of the independent variable. This stability cannot be assumed when there is a natural trend for the levels of the dependent variable to rise or fall even in the absence of the independent variable.*

EXAMPLE: A sample of depressed patients received a new anti-depressant medication. Most of the patients were much better six months later. Should we assume that the medication worked? No, most cases of depression are self-limiting or episodic: most patients snap out of it eventually (although they may relapse in the future). What is unclear in this case is whether those patients would have recovered even without the medication.

EXAMPLE: A sample of Alzheimer's patients received a new anti-confusional medication. None of the patients were any better six months later. Should we assume that the medication failed? Not necessarily, most cases of Alzheimer's progressively worsen over time. What is unclear in this case is whether those patients would have gotten even worse without the medication.

SOLUTION: These natural trends are dealt with better by a separate groups design. The sample would be composed of patients at the same stage of the disease and they are randomly assigned to one of two groups. The experimental group gets the new treatment and the control group gets a placebo. The natural course of the disease is present in both groups, so the difference between the groups will be attributable to the experimental treatment.

> **MISTAKE 36:** *Ignoring the impact of fatigue or boredom on repeated measures designs. These factors can greatly reduce performance on the subsequent measure. This is*

> *somewhat comparable to the previous case of a natural downward trend.*

EXAMPLE: Suppose I am trying to prove that Nikes are better than Reebocks. My operational definition of "better" is to run faster. So, I take ten people, have them put on Nikes, and then time them in a five mile race. As soon as the runners are back, I have them put on the Reebocks and have them run another five miles. If I observe that the runners did worse in the Reebocks, should I conclude that the Nikes helped them run better? No, they may have just gotten tired after running the first five miles.

EXAMPLE: Let's go back to the bare hand assembly performance versus that of the new glove. Suppose the task is extremely repetitive and boring so that the worker will probably lose motivation on subsequent trials. If he does ten trials with the bare hands, and ten more with the gloved hands and we observe gradually decreasing performance, should we conclude that the glove does not work? No, fatigue or boredom may have reduced the subjects' performance.

SOLUTION: One solution is to use a counterbalanced presentation. Half of the subject start with the bare hand, and then use the gloves; while the other half of the subjects start with the gloves, and then go bare handed. Have half of the subjects run the first time with Reebocks, and the other half run the first time with Nikes.

Another solution is simply to have two separate groups of workers: those assigned to the bare handed condition and those assigned to the gloved condition; or two separate groups of runners: those randomly assigned to Reebock and those randomly assigned to Nike.

Section Two: Errors in Sampling and Administration

Chapter 5: Is the sample good enough?

Too small? Or worth getting more?

If the number of subjects in the sample is too small, the results are not significant. Statistical significance is a formal term that means that we can be confident that random variation cannot account for our data.

> **MISTAKE 37: Assuming that significant data can be obtained from a small sample size.**

EXAMPLE: A company invested in some special training for its workers, and hypothesized that measured productivity would increase after training. Here were the before and after scores for each of the workers (n = 4)

Worker	Before	After
Adams	75	76
Baker	67	81
Carey	62	76
Davis	71	77

There is a difference in the means, there is not that much dispersion. Every subject improved, indeed, every after score was higher than every before score. Many novice researchers will jump to commit a Type I error, and declare a significant finding.

SOLUTION: Using a proper statistical test (in this case, a repeated measures t, a sign test, or a Wilcoxin) shows a lack of statistical significance due to small sample size. In general, the larger the sample size, the easier it is to attain statistical significance.

It is hard to come up with a rule of thumb for a minimal sample size (because the difference between the measures and variation within the measures must also be taken into account). However, I generally discourage researchers from doing one sample or repeated measures research if they have less than a dozen subjects, or less than twenty subjects for a separate groups or correlational design.

> *MISTAKE 38: Assuming that all of the returned questionnaires will be admitted into the final sample.*
>
> In order to control for confounding variables, it may be necessary to eliminate some of the subjects as being outside a targeted demographic. In a face to face questionnaire, or a group-administered questionnaire, the researcher can ask a few questions before giving the subject a form to fill out. With a mail out questionnaire, the researcher must put such information right on the questionnaire (or in some preliminary instructions) in order to determine if a given subject should be included in the final sample.

EXAMPLE: One student was handing out questionnaire to subjects as they exited the university library. After having deposited his filled out form in the ballot box, one subject said that everyone was so friendly at the university that he hoped he could attend some day after he got out of the military. The researcher had defined his population as university students, and had just assumed that everyone coming out of the library was a current university student at that institution. Now there was no way of knowing which completed form belonged to that one non-student.

EXAMPLE: One student was a secretary who wanted to survey parental attitudes. She decided to distribute a questionnaire where she worked. She apparently forgot that some of her colleagues were not mothers, until she was reminded by the comments on one completed questionnaire, "I don't have any kids."

70

SOLUTION: In a face to face questionnaire, or a group-administered questionnaire, the researcher can ask a few questions before giving the subject a form to fill out. With a mail out questionnaire, the researcher must put such information right on the questionnaire (or in some preliminary instructions) in order to determine if a given subject should be included in the final sample. The student research described above should have asked

Are you a student here?

before giving each subject a questionnaire. The secretary in the example above could have clarified in her instructions that the questionnaire was just for parents, or she could have had a background item on the questionnaire relating to parenthood.

> *MISTAKE 39: Assuming that all of the questionnaires matching the demographic profile of the sample will contain usable data.*
>
> You may get back 62 questionnaires all from your target population (e.g., students, nurses, police officers, accountants) but that does not guarantee that all of those questionnaires will be useable. Those questionnaires with incomplete or confusing responses will have to be "culled out" (eliminated from the tabulations done on the final sample).

EXAMPLE: One police explorer did a survey at the station, and her main hypothesis was that there were personality differences between sworn officers and civilian employees of the police department. Two questionnaires did not indicate whether the subject was a sworn officer or civilian employee. Another questionnaire failed to provide an answer on the question dealing with personality. Another circled halfway between two different responses on a personality question. All four of these questionnaires had to be eliminated from the tabulations done on that hypothesis.

SOLUTION: The best solution for the missing data problem is prevention. The fewer the questions, the more likely they will all be answered. The clearer the questions, the more likely they will be answered. The more definitive the response format (e.g., circling as opposed to checkmarks), the more likely the question is to be answered, and answered clearly.

It is important to note that some statistical programs used for analyzing data will automatically remove a subject from tabulation in a correlation matrix if there is any missing datum on any variable.

Representative enough?

MISTAKE 40: *Assuming that random means haphazard, or whatever comes by.*

The term random means equal probability. When there is random sampling that means that each member of the population had an equal chance (compared to the other members of the population) of being included in the final sample. The stark reality is that randomization is only assumed, and can never be proved. Wherever the researcher is more likely to include subjects who are more convenient to get into the sample, it is more correct to say that we have a "sample of convenience" rather than one that was randomly selected.

EXAMPLE: A news report of a mass murderer may claim that he started firing randomly at passers by. Actually, we can call it random only to the extent that all passers by had an equal chance to become targets. If the shooter tended to target tall or fat or male or white subjects, then it was not a random pattern.

EXAMPLE: A student stands outside of the university library and looks for other students who might fill out her questionnaire. She is a little shy and is reluctant to approach people with tattoos or who are smoking. She does not want to interrupt people who come out of the library already engaged in a conversation with someone else. She may not even be aware of how she is deciding whom to approach, but those decisions are going to lead to a sample which may over-represent introverts and Mormons.

SOLUTION: Utilizing a lottery system or a coin flip is the best way to truly say that the sample was selected in a random pattern. Otherwise, we should just admit that the sample was one of convenience, and we should explain our process of selecting specific subjects to participate in the questionnaire.

> **MISTAKE 41:** *Assuming that a larger sample is always a better sample. The quality of a sample is more dependent upon its representativeness than its mere size.*

EXAMPLE: A local newspaper published a national poll and ran these headlines. "Two-thirds of Americans support hand gun control." An irate reader fired off this letter to the editor. "So, a national sampling of just 1,232 Americans is used to determine that the American people want gun control. Why don't the pollsters come on down to the convention of National Rifle Association, where they can get the opinions of thousands of law-abiding Americans. I guarantee that the results will be different"!

SOLUTION: The reader was right in predicting that the results would be different, but wrong in assuming that a sample of individuals whose views reflect only one particular extreme on an issue would be a more representative sample of the broad range of views to be found in the entirety of the population of U.S. voters. Indeed, if the polling organization had wanted to save some money, and get a more impressively large sample size, they could have gone to the convention of the N.R.A. and just handed out the questionnaires. Fortunately, the polling organization had some integrity, and was interested in the quality of the data, not the quantity or the cost.

> **MISTAKE 42:** *Assuming that a larger sample is always worth the extra time and expense.*

EXAMPLE: Take the previous example about the polling organization that made phone calls around the nation to record the opinions of 1,232 American voters. Many novice researchers assume the following: "If we double our efforts, we will double our statistical significance."

SOLUTION: The only thing that we can be certain of by doubling our sample size is that certain costs in gathering the data will increase proportionately. There is a law of decreasing marginal returns that operates here. Is it better to have a sample size of 100 or 200? Obviously, adding an additional hundred subjects increases the chance of reaching a certain target level of statistical significance (e.g., a fair level of $p < .05$). Would adding another hundred (n = 300) improve things even more? Certainly, but the next hundred would not do as much to improve statistical significance as did the previous hundred.

So, how do we know where to stop? For doing a project for a class, I tell my students that 50 is about the right number because I would rather they spend more time analyzing the data and writing the report than in getting an addition fifty subjects to bring sample size to a hundred. An n of 100 is a better number for students writing a dissertation or trying to get an article published. National polling organizations (e.g., Gallup) usually get around a thousand respondents in order to have a two or three percent difference between Bush and Gore be 95% confident in the results ($p < .05$).

> ## MISTAKE 43: Trying to calculate a minimum sample size.
>
> There is a formula which will tell us the minimum required sample size if we can determine what level of significance (e.g., $p < .05$) we prefer and what is the dispersion (e.g., standard deviation) of the measured variable. Here are the steps involved.

1. Get the standard deviation of the population.

2. Decide what interval of approximation you require (e.g., within 10 points of the actual value).

3. Decide on what level of confidence you require (e.g., .05).

4. Look up the corresponding z-score (e.g., 1.96)

5. Multiply the standard deviation times the z-score, divide by the required interval, square the quotient.

6. If your answer is a decimal, round up to the next highest whole number.

The above approach works pretty well when we deal with populations for which we have known measures of central tendency and dispersion, and the dispersion is relatively small, and samples tend to approximate their populations in terms of central tendency and dispersion. In other words, this will work if the subjects are machine parts, but not if the subjects are voters, customers, workers or job applicants. Therefore, this approach is not very useful in estimated required samples for questionnaires.

EXAMPLE: One of my students wanted to do a study on the use of a particular pre-employment screening test at the company in which he worked. He looked up the norms in the accompanying test manual and found the mean and standard deviation based upon years of using the test in a variety of industries. He chose the interval he wanted (10 points) and the level of confidence (p < .05) and found the corresponding z score of 1.96. He crunched the numbers on his calculator and the required n was 34.12 subjects. He knew that he would have to round off to the nearest who person on this discrete scale. At first he assumed that he could just use the general rules for rounding off (and round down to 34) but then he re-read the specific instructions for this equation which said that he had to round up, and determined that he needed 35.

He handed out 35 questionnaires and obtained p = .07. He was so disappointed because he had assumed that following the formula for minimum n would guarantee him statistically significant results on every hypothesis he had initially advanced. I explained to him that following the minimum sample size rule merely made it unlikely that he would have to accept a null hypothesis just because of an excessively small sample size. He would still have to accept the null hypothesis on the vast majority of times (95% to be exact) if there were in fact no real difference between the groups.

Furthermore, I asked him if his assumption that the central tendency and dispersion of his sample was sufficiently close to the population norms to justify his initial decision on sample size. He went back and calculated the mean and standard deviation of his sample, and proudly announced that they were not significantly different from the population norms. "Now you are taking so much pride in finding a lack of difference," I mused. "Is it not possible that you are merely accepting the null due to an inadequate sample size?" "Oh no," he responded again with unwarranted pride, "remember, I began with the formula for getting an adequate sample size." "Yes, you did, but that assumed that you had the same mean and standard deviation in your sample, something which at that point, you could not prove, only assume."

This assumption is frequently unwarranted in the case of human subjects due to the difficulties of truly random sampling. In the above example, were the specific job applicants truly a random sample reflecting the ability and motivation levels of the nation's workers of the last thirty years? All of the applicants were applying for a job in a given industry (fast food) with a given company in a given geographical region.

SOLUTION: Do not assume that the formula for minimum sample size guarantees statistical significance or even attains a sufficiently large sample given the complexities of sampling human populations.

MISTAKE 44: Stopping sampling just because a minimum number has been reached.

EXAMPLE: One of my students took me too literally when I said that she only had to do about fifty subjects. She mailed out 150 figuring on a one-third response rate. She got back 62 completed questionnaires, she decided not to count the last twelve to come in. Most of these were from one site, and represented the majority of questionnaires filled out at that site.

EXAMPLE: Another of my students only made fifty photocopies of a questionnaire he was intended to give out at an AA meeting. There was a good turnout that evening and almost seventy showed up. He decided to distribute the questionnaires to those fifty who had the most interest in "voting" on the issue. "Since we do not have enough to go around, I am going to ask you if you do not have a strong opinion on this topic, then please let someone else fill out your questionnaire." This resulted in a sample of those who just had the most extreme opinions.

SOLUTION: There are sometimes good reasons for removing a questionnaire from the sample. One aspect of the analogy with an election is pertinent here. If someone is qualified to vote, and has filled out a readable ballot, that person's vote should count. Bigger samples are better than smaller samples (although always so much better that they are worth the extra effort to increase sample size). But here the extra effort is in reducing the sample size, and that effort is never justifiable by size alone.

The above examples represent exceptionally bad sampling because the way in which the samples were reduced also made the samples less representative. The first student disproportionately disenfranchised one particular site. The second student's decision to give preference to those subjects with stronger opinions on the topic shifted the sample's stance on the dependent variable away from the moderate position.

MISTAKE 45: Failure to identify the target population within the available sample.

EXAMPLE: You want to find out if those Carl's Junior drippy hamburger ads are effective with young people. So, you develop a questionnaire to see if people can at least identify the ad as being Carl's Junior. Then, you distribute a questionnaire at a private southern California university and find that most of the students cannot identify the sponsor, and worse yet, they found the ad to be "gross" and hit the channel flipper whenever it comes on.

Before you conclude that the ad has been a failure, we should identify the university from which the sample was obtained as Loma Linda University, which is affiliated with the Seventh Day Adventist Church (most of whose members follow the recommended vegetarian diet). Vegetarians are not the target market. The real question is whether the ad is effective with people who like hamburgers.

EXAMPLE: You want to see if people under 21 favor changing the legal drinking age to 18. You distribute a questionnaire during the day time at a community college (where most but not all) of the students are over 18 but under 21.

EXAMPLE: You want to measure the attitudes about the local public schools held by the parents of children in those schools. You have an opportunity to survey about two dozen parents at a local soccer game between two local teams of 9 year olds. One problem is that not all the parents have children in public schools. Some of the children might be in private schools (or even home schooled).

SOLUTION: One solution is to draw the sample from a population known to be the target population. If you are in a college cafeteria, give the questionnaire to those students eating hamburgers.

When you cannot be assured that your sample consists exclusively of the target population, put in some questions that will later help you identify the target. Ask the community college students about their age, and only include those over 18 and under 21 in the final sample. Ask the parents at the soccer game if they have any children in the local public schools.

Grouping: experiment? Quasi-experiment? Survey?

> *MISTAKE 46: Getting two comparison groups from very different populations so that many confounding variables are introduced.*

EXAMPLE: One student at a women's college was trying to show that women were more subject to depression than men were. It was easy to get a sample of women from the college students, but for the men she decided to ask the male professors. Significant differences were noted, but were they due to gender? or to age? or to occupational status?

SOLUTION: Gather the comparison groups from the same population, or in such as way as to minimize the presence of confounding variables. In the example above, the investigator should have looked for a group of male college students: comparable in age and other background factors, so that only gender would distinguish between the two groups.

> *MISTAKE 47: Allowing subjects to choose their own grouping in an experiment, means that it is no longer an experiment but a correlational survey with two dependent variables.*
>
> The essence of an experiment is that an independent variable must be manipulated. In a separate groups design, this is best accomplished by randomly assigning subjects to one of two groups, and then treating those two groups differently. When the grouping is determined by the subject's own choice or preference, then grouping reflects a dependent variable, not an independent variable.

EXAMPLE: Suppose I wanted to show that watching violent TV programming promoted subsequent violent behavior. So, I arranged two different rooms: one showing something non-violent (e.g., Barney the Dinosaur) while the other room had violent programming (e.g., Jerry Springer's out takes that were too violent for broadcast TV). Suppose I do not want to force my subjects into watching something that they do not like, so I tell them all "You were nice enough to volunteer for this research, I am going to be nice enough to allow you to choose what you want to watch." Each subject then chooses whether to watch violent or non-violent TV, and then over the next week we measure the level of violent behavior of each subject and we find that those who saw the violent TV act out more violence.

Should we conclude that the violent TV caused the subsequent violence? No, another reasonable conclusion could be that violent people prefer violent TV.

SOLUTION: The best way to infer cause and effect is to do an experiment in which an independent variable is manipulated, and that means random assignment, not self-selected grouping by the subjects.

MISTAKE 48: Assuming random assignment within an experiment, when the groupings were predetermined.

EXAMPLE: A psychologist wanted to investigate the impact of listening to a call-in radio show on the listener's reported levels of anxiety. She went into each of two different early morning psychology classes (one met on Monday-Wednesday-Fridays, and the other on Tuesday-Thursdays). On Monday morning, she played a tape of Dr. David Viscott (a radio psychiatrist) who invites calls about emotional problems, romance, and parent-child relations. For the Tuesday class she played a tape of Dr. Dean Edell (a radio physician) who invites calls about a broad range of physical problems. The first group showed slightly lower levels of anxiety compared to that of the second group.

Does this show that listening to the psychiatrist helped them (at least temporarily)? That conclusion is only possible if we can assume that the two groups started out with roughly equal levels of anxiety. It is also possible that one group of neurotic friends decided to sign up for the Tuesday class while a better adjusted group of friends happened to have signed up for the Monday class.

SOLUTION: A true experiment requires random assignment. That would mean that each member of the sample has an equal chance (compared to every other member) of being assigned to one group. This does not mean that the two groups must be of equal size. A sample size of fifty may have random assignment even if the experimental group only has ten and the control group has forty: the important thing is that each one of those fifty had an equal chance of making it into that experimental group. The best way to assure this randomization would be to make assignments by means of a lottery or coin flip.

In the absence of random assignment, as in the above example about the radio talk shows, this cannot be called a true experiment, but a quasi-experiment. The best way to factor in the impact of the initial grouping on the dependent variable would be to add a repeated measures dimension: in each group measure anxiety before and after the tapes are played.

Ethical compromises and sampling.

Social science research must be guided by several ethical principles. With all both human and animal research subjects there is a need to minimize the potential harmful consequences of the research: risks, dangers, and suffering. This is one reason why many topics in psychology (e.g., loss of parents at an early age, brain damage) cannot be studied by experiments on human subjects. These topics can only be studied by surveys (which are not as good at eliminating other possible independent variables) or by experiments on animals. Indeed, one of the most controversial areas of scientific research is the justifiability of the pain (and even death) given to the more than twenty million animal subjects each year in U.S. laboratories.

With human subjects, there are guidelines about confidentiality and the anonymity of the subjects. Someone reading a case study of a patient should not be able to infer that the 56 year old male divorced accountant is really Mr. Jones down the street. Many hospitals, companies, and schools have internal review boards or written policies which limit what kinds of data can be used in research. So, before you start distributing your questionnaire around the office, or sifting through the employment records, make sure that your research is permitted.

Another issue with human subjects is that of informed consent. Each participant should have a clear idea of what he or she is getting into upon agreeing to fill out a questionnaire or be subjected to the manipulation of the independent variable. One great limitation of this guideline is that many of the subjects that psychology is most interested in (e.g., children, the retarded, the mentally ill) might be incapable of fully comprehending the ramifications of their participation in the research.

An additional constraint on informed consent is that certain experiments necessarily involve some degree of deception of subjects because if they knew what the research was really about, that would affect the way that they responded. It may be necessary to lead subjects to think that the research is about one variable (e.g., attitudes about the qualifications of job applicants) when actually it is about another variable (e.g., discrimination based upon the gender or ethnicity of the job applicants).

A related guideline is debriefing. After the research has been completed, subjects should have the opportunity to learn about the results. If the subject has experienced any stress due to the research process, he or she should have an opportunity to get some counseling in order to overcome those problems.

Another guideline is freedom from coercion. Subjects should not be forced to participate in the research. Prisoners should not be promised special consideration in parole hearings for their agreement to participate in dangerous experiments. Even students in psychology courses should not be required to participate in research as part of their grades, but should be given alternative ways of meeting course requirements.

While most of these ethical concerns apply to experiments rather than questionnaires, and to the clinical aspects of research rather than that done on most organizational surveys, many sampling issues may compromise ethical guidelines of research.

MISTAKE 49: Coercing subjects to participate in a questionnaire.

EXAMPLE: Employees in a secretarial pool were told in a meeting that their cooperation in filling out a questionnaire was expected. Not only did this violate the principle of non-coercion, but any attempt to enforce compliance by tracking who had turned in the questionnaire and who had not could also violate the principle of anonymity of the subjects.

SOLUTION: Announce that filling out a questionnaire is voluntary, but convey to the subjects that the resulting data will be useful for the organization and will in no way compromise the anonymity of the subjects.

84

> *MISTAKE 50: Having a background question (or series of questions) which subjects might fear as compromising their anonymity.*

EXAMPLE: Consider what might happen if this questionnaire were administered in a small real estate office.

What is your gender?

MALE   FEMALE

How old are you?

UNDER 30   30-39   40-49   OVER 50

What is your job title?

SECRETARY   SALES

Assuming no layoffs, how likely is it that you will be working for this office at this time next year?

VERY LIKELY   SOMEWHAT LIKELY   NOT VERY LIKELY

Imagine the concern of a 53 year old secretary as she realizes that she is the only secretary over age 50; or a 29 year old salesman, who can think of only one younger sales person (a woman). These workers know that the three background variables create a pattern of identification as unique as their fingerprints. The result is that these workers are discouraged from expressing opinions that might get them into trouble. In this case, these identifiable workers may be less willing to admit their intentions to seek employment elsewhere.

SOLUTION: Adjust the measure of background variables so that subjects will not think it possible that their identifies can be traced from a specific pattern of answers. For example, the 53 year old secretary's anonymity might be preserved if the age range were changed to 40+ (so that a few additional secretaries might be included).

Chapter 6:
Administering the questionnaire

The administration of a questionnaire is the process of getting it out to and back from the subjects. Many problems in ethics or in sampling result from inappropriate administration of the questionnaire. Each approach to administration involves certain limitations and risks.

Group administration

> *MISTAKE 51: If subjects are approached in groups, there is a risk that they will look at each other's answers, compromising their anonymity, or even leading to collaborative answers.*

EXAMPLE: Back in 1983 one of my students was distributing a questionnaire measuring student attitudes about the proposal for a nuclear freeze. This was an extremely controversial topic on our campus at the time. The potential subjects were approached outside of the library and asked if they would participate in a survey about the nuclear freeze. While most students were alone, some were in groups of two or more. As I walked by, I noticed a couple of subjects debating the issue as they were filling out the questionnaires. One more passionate student was trying to convince her friend, "Don't vote that way, before you know all the facts."

SOLUTION: One solution is to only approach unaccompanied persons. However, this could lead to a non-representative sample if the topic dealt with relationships. Another tactic would be to give oral or written instructions that subjects should fill out the questionnaires without conferring with anyone else.

Face-to-face, one-on-one

> **MISTAKE 52: *Ignoring the impact of the presence of the observer on the gathering of data.***

EXAMPLE: In the classic Hawthorne electric works study of the 1920s, workers greatly increased production whether lighting was increased or decreased. One of the accepted explanations for these paradoxical data has become known as the "Hawthorne Effect": that the mere presence of the observers with their clip boards, white coats, and stop watches motivated the workers to increase productivity.

EXAMPLE: The owner of a fashion boutique in Caracas thought that she would emphasize those fashions that attractive men preferred women to wear. So, she administered a questionnaire face-to-face to a sample of good looking men coming out of a subway station. To her amazement, the men seemed to prefer one kind of dress one day, and a very different outfit the next day. Then she realized that what they men said that they preferred was whatever outfit that she herself had been wearing that day. This is known as "Demand Effect" or "Courtesy Bias."

EXAMPLE: About fifteen years ago, the U.S. Congress was debating whether Japanese-Americans who had been forcibly interned in relocation camps during World War II should be paid compensation for this injustice. Two students developed a questionnaire that measured the subject's support for a proposed bill involving compensation, as well as the subjects' background variables of age, gender, and ethnicity.

What made this research an experiment was the stimulus provided by the ethnicity of the researchers. Both of the student researchers were women in their forties, but one was a blond Caucasian and the other was Japanese-American (who had actually been in one of those camps during World War II). The demand effect was profound in this case: only about half of the subjects who received the questionnaire from the white woman advocated compensation, but all of the subjects who received the questionnaire from the Japanese-American woman advocated compensation.

EXAMPLE: In the wake of the first trial of the Los Angeles police officers who had been accused of beating Rodney King, two of my students distributed a questionnaire asking subjects if they agreed with the Simi Valley jury's acquittal verdict. There were other items on the questionnaire measuring background variables of the subjects: gender, age, ethnicity. Subjects probably assumed that the purpose of the research was to correlate ethnicity with acceptance of the verdict. This was such an emotionally charged issue that the verdict allowing the white police officers to walk free was seen as a great miscarriage of justice by many African-Americans, and sparked several days of rioting.

Actually, this was an experiment. There were two students doing the research and handing out the questionnaires. Both were women in their forties. One woman was a blond of northern European extraction, while the other was African-American. The initial hypothesis was that subjects receiving the questionnaire from the African-American woman would be less likely to express acceptance with the acquittal verdict. While some of the subjects supported the verdict regardless of who gave them the questionnaire, subjects were less likely to support the verdict if they had received the questionnaire from the African-American woman.

SOLUTION: Realize that the presence and characteristics of the person who is asking the questions or making the measurements may very well influence subject behavior. One possible solution is to make the observer less visible (e.g., observation from behind a one-way mirror). Another solution is to neutralize the impact of the person handing out the questionnaire. The Venezuelan model could have worn a skirt (instead of a dress or pants). The question about the Rodney King trial (in which the victim was Black and the accused police officers were White) could have been handed out by someone of Asian ethnicity.

> *MISTAKE 53: Questionnaire distribution and collection procedures that fail to preserve the anonymity of the subjects.*

EXAMPLE: The very first questionnaire I ever developed was back when I was a student in an introductory psychology class at Claremont Men's College in 1967. (It is now known as Claremont McKenna College because it has gone co-ed.) Our subjects were the students who attended the two adjoining women's colleges: Pitzer (which has since gone co-ed) and Scripps (which remains a woman's college). We had hypothesized several differences between the two colleges and had items dealing with a variety of traits and behaviors (including sex and alcohol). In retrospect, one great problem with that survey is that we had done nothing to safeguard the anonymity of our subjects. We approached individual women, convinced them to do the questionnaire (which took less than five minutes) and then the women handed the questionnaires back to us.

Some of us could not resist the temptation to immediately look over the answer sheet of an attractive female as soon as she walked away in order to answer our personal question "Would she be fun at a party"? Even if none of us had peeked at the answers and associated that with the girl who had just handed us the filled out form, there is still the risk that the female subjects might have worried that this lapse of propriety on the part of the researchers was possible. This may have led some subjects to understate their participation in some embarrassing activities in order to avoid being labeled.

SOLUTION: Before asking the subjects to fill out a questionnaire, show them the ballot box, and tell them that they are not to hand the completed questionnaire back to you, but that they are to fold the questionnaire and then place it in the ballot box. This assures them that individual questionnaires will not be associated with individual subjects.

MISTAKE 54: Failure to mask embarrassing answers when a ballot box cannot be used.

EXAMPLE: A telephone survey attempted to investigate safe sex practices. Even though subjects were told that the data would be kept in strict confidence, many subjects were frightened and denied that they had engaged in any unsafe sexual practices.

SOLUTION: One approach is to give the subject another possible reason for giving a yes answer. Consider these instructions.

> For the next question, I want you to take out a coin and flip it right now. Tell me when the coin has been flipped, but do not tell me yet if the coin came up heads or tails.
>
> Now listen to the following question, but do not give me a yes or now answer.
>
> During the last twelve months have you had unprotected vaginal or anal intercourse with at least two different partners?
>
> If the coin came up heads, OR if you have had vaginal or anal intercourse with at least two different partners, say the word "heads." In other words, you will say "tails," only if the coin came up tails AND you have not have vaginal or anal intercourse with two or more partners.

In that way, a "heads" answer does not prove that the subject admits to unsafe sex, and there is no way of knowing if a given subject has or has not had unsafe sex, however the aggregate frequency of unsafe sex can be estimated by subtracting .5 from the proportion of "heads" answers and then doubling that number.

> **MISTAKE 55: Filling out questionnaires using a face-to-face approach can lead to subjects understating embarrassing or unpopular positions.**

EXAMPLE: In the 1982 gubernatorial election in California, exit polls were used. As voters left the polling place after having cast their ballots, reporters with microphones and cameras asked them who they had voted for. These exit polls indicated a narrow victory for Democrat Tom Bradley, the mayor of Los Angeles (who happened to be African-American). The subsequent counts of the actual ballots cast in the polls that day showed a narrow victory for the Republican, George Deukmejian. Some voters who had actually cast a ballot for Deukmejian had told the interviewers that they had voted for Bradley. The probable reason was that they were afraid that not voting for Bradley would be seen as racist. The number of voters described by this scenario may be small, but it was large enough to flip the results from a narrow Deukmejian majority to a narrow predicted victory for Bradley.

SOLUTION: Using ballot boxes would be the obvious solution. What most of the networks then went to after this election was a focus on quick counts of actual ballots right after the polls closed. A few key precincts would be a sufficient sample if they could be identified as "bellweathers" which historically had outcomes that were similar to those of the state wide aggregates.

Telephone

> **MISTAKE 56: Assuming that telephone surveys yield a random sample.**

EXAMPLE: You know that in some metropolitan areas, the proportion of unlisted numbers approaches 50%, so you get a random digit dialer going from 5 to 9 pm on a weeknight. You reach some government offices and businesses (most of which have closed by that time), but you are able to reach more than a dozen people in your target population each hour. The questionnaire is brief and covers an interesting topic, so your refusal rate is quite low. You will get a large sample easily and quickly, but should you assume that it is a random sample of the population?

Before World War II, the answer would have been no, because telephones were more prominent among the wealthy than the poor. Today, most households at every income level have a phone, but they may not have an equal probability of having someone who will answer that phone. Younger adults without children will be less likely to be home, more likely to be working, exercising, taking a night class, or out on a date. Wealthier people will be likely to screen out unwanted incoming calls with an answering machine, caller-ID, or even a live butler.

SOLUTION: Go ahead and use telephone surveys, but just realize that they under-represent certain segments.

> **MISTAKE 57: Attempting oral administration of a multiple nominal scale.**
>
> When the number of categories are large, or initially unknown to the subject, there is a difficulty of keeping them all in mind at the point of making a decision on just

> one. (This problem is compounded when the subject is young, retarded, or ill.) The same problem might exist, but to a lesser degree with ordinal scales involving many levels.

EXAMPLE: When we tried to administer the Beck Depression Inventory orally to nursing home patients, we had a great deal of difficulty. Each of the questions had four possible ordinal responses (and some of them quite long).

Many of the subjects would forget some of the items before deciding on a specific answer. Instead of selecting one specific answer, some subjects just started talking in general about depression, not leaving us with any scoreable response.

SOLUTION: One solution used in face-to-face polling is to provide the subject with a cue card showing the specific acceptable answers of a multiple nominal scale. For telephone administration, the number of possible answers must be kept to three or four in most cases.

Regarding the problem of oral assessment of geriatric depression, we decided not to use the Beck Scale or others (e.g., Zung, Hamilton, Center for Epidemiological Studies) that made use of ordinal or multiple nominal scaling. We constructed the Geriatric Depression Scale so that it would have a simple "yes/no" format because we were convinced that more severely depressed patients would be taking the scale orally.

> **MISTAKE 58: Using words which are frequently misunderstood due to their pronunciation.**
>
> Especially over the telephone, subjects may readily mistake one word for another. This is especially a consideration with elders and other hard of hearing subjects.

EXAMPLE:

Should homosexual marriages be outlawed?

STRONGLY AGREE   AGREE   DISAGREE   STRONGLY DISAGREE

Some of the subjects who heard the question read over the telephone heard the last word of the stem as "allowed" which completely reversed the answers.

SOLUTION: Use a pilot survey to make sure that the questions can be clearly understood by most subjects.

Mailouts

> **MISTAKE 59:** *Assuming that all (or even most) of distributed questionnaires will be returned. The response rate on mailed out questionnaires can be quite low, and is always unpredictable.*

EXAMPLE: One student was a police explorer. She distributed a questionnaire in the mail boxes at the police station so that each of the 65 sworn officers in the department got a copy. Less than half filled out the questionnaire.

She had handed out the survey during a summer month: some officers were gone on vacation, others were off on reserve military duty, some had been dispatched to help with a forest fire, and those who remained were working very long hours. Few of them could be bothered with filling out a questionnaire.

SOLUTION: Expect a low response rate. The rate can be improved by several tactics. One is to keep the questionnaire short. Another is to make sure that it does not hit the subjects about the same time as some other major event or responsibility competes for their attention. (Do not survey accountants during tax season, do not survey professors at the beginning or end of a semester).

> **MISTAKE 60:** *Failure to set a return deadline on mailed out questionnaires*
>
> *If there is no date, some subjects will put it in a large pile of papers that they are sure to get around to. The problem is, they may not get around to it until after you have finished your report.*

EXAMPLE: One of my students was a teacher. She distributed questionnaires in the mail boxes of each of her colleagues. She did not put a return date on the questionnaire. Very few had come in by the time she had to go on the computer and analyze her data. After she wrote the report, a majority of the questionnaires came in.

SOLUTION: Print a clear due date on the questionnaire. The date must be early enough to give you sufficient time to analyze your data and write your report. Make it late enough to give your subjects enough time to do it, but not so late that they will put it in one of those "to do piles." The optimal situation is that the subject will look at a short questionnaire, and then decide to fill it out right there in the mailroom without even taking back home or to the office.

MISTAKE 61: Constantly revising the sample to include late questionnaires.

EXAMPLE: One student did not get her report in on time because after she ran the numbers on the computer the first time, five more questionnaires came back in the mail, and she was hoping for statistical significance, and thought that a slightly larger sample would help. She ran the numbers again but got even a smaller difference between the groups. Then a couple of days later three more questionnaires came back and so she crunched the numbers again, and still failed to attain statistical significance, and realizing that she was up against my rigid deadline for her first written draft, gave up on writing the report.

SOLUTION: The election analogy is helpful here too. The polls close at a certain hour. Someone cannot show up the next day, after the votes have been counted and the winner announced, and request a ballot. It is only practical to stop at a certain point and deal with the questionnaires which have arrived. What was particularly objectionable about the above student's approach is that it was reminiscent of Al Gore's post election strategy: keep on counting until we get the desired results.

MISTAKE 62: Failure to include information on where to return a questionnaire.

> You would be surprised to find out how often this simple mistake is made.

EXAMPLE: One student was a nurse in a large hospital. She put a questionnaire in each of her colleague's mail boxes. A couple of days later, someone who knew that she was distributing the questionnaire brought to her attention that there was no identification of who the researcher was and how the subjects were supposed to return the completed questionnaires. The nurse conducting the research then had to send out another announcement in everyone's box in which she identified herself as the researcher and then explained how to get the questionnaires back to her. Then another colleague complained that he had thrown away the first questionnaire because it had no return information, so the researcher decided to make enough copies to put in everyone's box again, just to make sure that everyone would have a copy. After she analyzed the returned questionnaires and wrote the report, another of her colleagues complained that he resented having to "vote twice" on the matter (he had filled out both questionnaires and returned them to the researcher). She had no way of knowing how many of her colleagues had done this, and had no way of recognizing which questionnaires were duplicates.

EXAMPLE: An intern at an accounting firm distributed a questionnaire to the accountants. She did identify herself, but did not clarify how they were supposed to get the filled out questionnaire back to her. Most of the accountants just put the filled out questionnaire in her mailbox there at the accounting firm, but he internship was over that week. What she had hoped is that the completed questionnaires would be returned to her university address.

SOLUTION: Put the instructions on how to return the questionnaire right on the questionnaire (preferably at the beginning AND the end). If you only put it on a "cover letter" this runs the risk that the two pieces of paper will be separated. Another good idea is that when postal mail is involved, a stamped self-addressed envelope can be used.

MISTAKE 63: Not building motivation for the subjects to return a mailed out questionnaire.

EXAMPLE: A former police explorer distributed a questionnaire at the local police station, putting one in the box of each sworn officer. He did not mention on the questionnaire that he was a former police explorer who had worked with another department. Some of the officers remarked that they did not have the time to help just another college kid.

SOLUTION: One often used technique for building motivation is to offer the subjects the chance to see the results of the survey. I think that an even better approach is for the researcher to identify with the subjects. This is a good way to begin most of the questionnaires distributed in the office mail:

Dear colleague, you know me as Joan Smith, R.N., in I.C.U. What you may not know is that I am also a student at the University of Redlands, working on a bachelor's degree in management. You can help me with a project I am doing by filling out this questionnaire and putting it in my mailbox by next Friday, May 17, by 5:00 PM. Thanks, Joan.

Email and web-site

Email and web-site surveys involve problems with the recording of responses (which will be discussed in Chapter 8) but they also involve problems in sampling.

> ## MISTAKE 64: Assuming that mail out or web surveys yield a random sample.
>
> These types of surveys require a higher level of motivation in order to get a subject's response. "Non response" rates tend to be larger than they are for face-to-face or even telephone surveys.

EXAMPLE: A student put a questionnaire on abortion in the mail boxes of each of the other students in her dorm.

Do you think that abortion should be ...

MADE ILLEGAL UNDER MOST CIRCUMSTANCES

HANDLED PRETTY MUCH THE WAY IT IS NOW

MADE LEGAL UNDER ALL CIRCUMSTANCES

Less than half of the students in her dorm responded. To her amazement, few wanted to support the status quo and large numbers selected the extreme answers. The subjects with the more moderate positions were less motivated to respond.

EXAMPLE: People received an email inviting them to click on a link and take a survey on the web. Of almost a thousand email invitations sent out, less than a hundred people went to the web link, and less than fifty completed the questionnaire on the web. The questionnaire dealt with attitudes about the legalization of drugs. Most of the completed questionnaires favored the legalization.

Should we conclude that most Americans prefer legalization? Look at the great chain of assumptions that must be met in order to assume that we have a representative sample. First, we assume that people who have an email address are representative of all Americans. (Actually, we know that they tend to be disproportionately young, white, Asian, and/or educated). Second, we have to assume that the people on this email list were representative of all persons online. Unless the list was generated in a truly random fashion (which would require the undesirability of spamming large numbers of persons not related to the researcher), the answer is probably no. The list is probably composed of persons related to the researcher. Third, we have to assume that the 10% of the people who chose to go onto the questionnaire link are representative of all the persons who received the email. They would probably disproportionately represent people who have strong views about drugs. Fourth, we have to assume that the people who choose to stay on the site and complete the questionnaire are representative of all person who initially went to the site, while they were obviously those who were the most motivated to invest the time it would take to express an opinion.

SOLUTION: Use techniques to increase response rate of mail order surveys. Alternatively, use this sampling technique to prove that there is a difference between a web sample and norms established by other techniques.

Branching questions

> **MISTAKE 65: *Using branching questions on a paper and pencil questionnaire.***
>
> These get respondents confused and many will not fill out the rest of the questionnaire correctly (or stop). Also, having branching questions usually means that some questions are to be skipped over, and this violates a basic principle that each subject should answer each question.

EXAMPLE: Consider this series of questions.

7. Do you have children?

YES   NO

(If YES, go to question#8, if NO go to question #21)

8. Do you have children of school age?

YES   NO

(If YES, go to question#9, if NO go to question #21)

9. Do you have any children in a private school?

YES   NO

(If YES, go to question#10, if NO go to question #21)

10. Do you have any children in a public school?

YES   NO

SOLUTION: If branching questions are required, it is best to use a face-to-face or computerized administration of the questions because these approaches allows the person or program administering the questionnaire to decide which questions to ask and simply skip over those not to be asked, and not confuse the subject with them. If the questionnaire must be administered on paper, it is best to collapse branches into one multiple nominal response format. Consider the above series of four questions. In general, put what is likely to be the most frequently selected response first.

What kind of school do your children attend?

> I HAVE NO CHILDREN
> I HAVE CHILDREN, BUT NONE ARE OF SCHOOL AGE
> I ONLY HAVE CHILDREN IN PUBLIC SCHOOL(S)
> I ONLY HAVE CHILDREN IN PRIVATE SCHOOL(S)
> I HAVE CHILDREN IN BOTH PUBLIC AND PRIVATE SCHOOLS
> I NOW "HOME SCHOOL" ALL OF MY CHILDREN

Dangers of creating expectations

> *MISTAKE 66: Not realizing that measuring an organization's attitudinal climate changes that climate by creating expectations.*

EXAMPLE: Asking workers what they would like in terms of fringe benefits focuses their interest on that area and builds an expectation that something will be done. They might say a few weeks later: "When do we get those better medical benefits, the ones we voted for in that election?" Even an informal focus group can build such an expectation.

SOLUTION: Don't measure worker preferences unless there is some possibility that the worker's desire can be translated into company policy. When conducting focus groups, make it clear that the researcher is just gathering information, and has no power to bring about desired changes.

Excessive length

> **MISTAKE 67:** *A questionnaire that is very long, covering several pages, especially when part of one item is on one page (e.g., the instructions or description of a scenario) and the rest of the item (e.g., answers) is on the next page.*

EXAMPLE: One of the worst offenders in this category was one questionnaire about pickup trucks. For each question the subject was supposed to select one particular make of truck, e.g., "Who makes the most durable small truck?"

"Who makes the best looking small truck?" The subject was supposed to respond to each question using a fill-in-the-bubble answer sheet. There were five columns: A, B, C, D, E. The first page of the questionnaire gave the coding: A = Ford, B = Toyota, C = Dodge, D = Chevrolet, E = Nissan. Although the questions spilled over onto at least three pages, the coding instructions did not, and the subjects could easily forget. Some even made the assumption that C must mean Chevy and D must mean Dodge.

SOLUTION: The first suggestion is to have a questionnaire that is short enough to fit on one page. The second suggestion would be to put all essential information on each page. The third suggestion would be to customize the fill-in-the-bubble answer sheet so that it had the scoring coding right on top.

> **MISTAKE 68:** *Using a two-page questionnaire that is printed on one piece of paper, back to back.*

EXAMPLE: One medical team was doing a study of depression. The questionnaire involved about patient ten background items (e.g., age, gender, previous medical conditions) and the thirty item Geriatric Depression Scale. If all of the items were to be on one side of one sheet of paper, they would have been too small for some of the older patients to read. So, it was decided to use both sides of the paper, twenty items on each side. Unfortunately, nearly a fifth of the questionnaires which were collected were filled out on the front side only: the patients answered the tenth depression question, and assumed that they were done with the questionnaire, and handed it in, leaving most of the questions (all of those on the other side) blank. Therefore, these questionnaires did not have scorable GDS scores.

SOLUTION: The first suggestion is to keep the questionnaire short enough so that it fits on just one side of one piece of paper. In the above example, this would not be possible, because the GDS has a standard length of 30 items.

The second suggestion is to put at the bottom of the first page some instructions such as "Please complete the other side of this form" or "Please continue on next page." It helps to use a font different from that of the questions and answers. Highlighting these instructions in yellow also increases compliance.

A third suggestion is to only print the questionnaire on one side of each piece of paper, so that a two page questionnaire has two pieces of paper stapled together. This makes the existence of the second page obvious, but it is more work for the person who had to analyze the data later.

MISTAKE 69: Not stapling together separate sheets of paper.

EXAMPLE: Another medical team had a long patient questionnaire running to four pages. Running out of staples, paper clips were employed. Some of the paper clips came off as the completed forms were placed into the ballot box or being removed from the ballot box and prepared for analysis. Unfortunately, since all of the second pages look like all the other second pages, and were completely anonymous, it was impossible to tell which second page was associated with which first or third page. In this example, the GDS items were spread across two pages, so it was impossible to assign GDS scores for these separate questionnaires. Even if all the GDS items had been on one page (and a score could have been assigned) those scores could not have been matched with background data on the first page in order to test any of the hypotheses about which kinds of patients were more depressed.

SOLUTION: Use staples, not paper clips.

> **MISTAKE 70: An online questionnaire has so much material on one screen that the subject must scroll in order to see vital information.**

EXAMPLE: One marketing research company had a questionnaire that measured the consumers' perception about different brands of beer. The questionnaire was arranged in rows and columns. On the left was a short, descriptive phrase; on the top was the brand of beer: so each row was a different attribute, and each column was a different brand. There were at least a dozen attributes and at least eight brands of beer. There was no way that my screen could see all of the brands, or all of the attributes on one screen without scrolling.

	Miller	Budweiser	Corona
Refreshing			
Expensive			
Young			

One risk is that the subject will assume that he is done, when he comes to the end, not realizing that he has to scroll to the DOWN the page to see more attributes (rows).

Another risk is that the subject will assume that only those alternatives (columns) which appear on the screen are being measured, not realizing that he has to scroll RIGHT to see more possible examples of brands of beer.

The greatest risk is that the subject will scroll down to see the remaining attributes (or to the right to see the remaining alternatives) and then not remember what each row (or column) referred to, and end up clicking the wrong button.

SOLUTION: Put each item on one screen. Have that screen contain the instructions, the question, and all possible responses. The downside is that you have to make sure that there is rapid movement from one screen to another. Field test your design on several different screens (e.g., Mac vs. PC, Juno web with ad banners) in order to verify that everything is coming through on one screen without scrolling.

Section Three: Errors in Phrasing

Chapter 7: stem patterns to avoid

Vagueness

MISTAKE 71: *A vaguely worded question*

EXAMPLE:

How important is it to you that your next car be a real value?

As beauty is in the mind of the beholder, so value is in the mind of the consumer. Value could mean economical gas mileage to one car buyer, and affordable luxury to another.

SOLUTION: It is better to measure the specific components of value.

How important is it to you that your next car have excellent gas mileage?
EXTREMELY   VERY   SOMEWHAT   SLIGHTLY   NOT AT ALL

MISTAKE 72: *Using terms which the researcher thinks of generically but the subject interprets specifically. This results in an under-reporting of the level of the variable.*

EXAMPLE:

How long have you held your current job?

"I have been a Secretary Level Two here for only a few months. I was a Secretary Level One for five years here before that."

EXAMPLE:

Do you own a car?

"No, I own a pickup truck."

"No, I do not own. I only lease."

EXAMPLE: The researcher may be looking to see which manufacturer of the car, and the subject might think which brand of car. General Motors makes GMC Trucks, Saturn, Geo, Cadillac, Pontiac, Buick, Oldsmobile and Chevrolet. Ford also makes Mercury and Lincoln, some Mazda models, and has recently purchased Jaguar. Chrysler also makes Dodge and Plymouth, and owns Jeep and Eagle.

> What kind of automobile do you drive?
>
> GENERAL MOTORS   FORD   CHRYSLER

"None of the above, I drive a Lincoln."

EXAMPLE:

> Did you buy a new house in the past year?

"No, I just bought a condo."

"No, the house I bought was pre-owned."

EXAMPLE:

> Did you attend church in the last seven days?

"No, I attend a Jewish synagogue."

SOLUTION: Use terms that are sufficiently encompassing for all answers that you want to identify.

> How long have you been working with this company?
>
> Do you own or lease a four wheeled motor vehicle?
>
> Did you attend religious services in the past seven days?

> **MISTAKE 73: Using terms which the researcher thinks of specifically but the subject interprets generically. This results in an over-reporting of the level of the variable.**

EXAMPLE:

> Did you get a meal in a restaurant in the past seven days?

"Yes, I picked up an order of fries and a milkshake at McDonald's."

EXAMPLE:

> Did you move in the past year?

"Yes, I went off to college to live in the dorm in September and then moved back home in December."

Spending time in a dorm, or a hotel room, is not the kind of move that the researcher is looking for.

> Did you buy a house during the last year?

"Yes, we got a mountain cabin for summer weekends out of the city."

"Yes, we bought a rental as an investment."

The question is alright for measuring sales of real estate but will over-report changes of residence.

SOLUTION: Use terms that are sufficiently specific for what you want to identify.

> Did eat a meal in a restaurant (with menu and waiter) during the last seven days?
>
> Did you change your principal residence in the last twelve months?

Notice how these re-phrasings change the previous "yes" answers to "no."

MISTAKE 74: Failure to state a time frame.

EXAMPLE:

> Have you driven after consuming more than two drinks?
>
> How much time after? within the same hour?
>
> Did you vote for president?

"Yes, I voted for Reagan, the first time he ran."

> Have you gotten a speeding ticket?

I would have to answer "yes" because I got one once when I was 20, but have not had one in over 33 years, but some 17 year old who has just gotten his license, and has not yet been caught, would answer "no."

SOLUTION: Clarify the time frame to be investigated.

> Have you ever driven within an hour after consuming two drinks?
>
> Did you vote for president in the 2000 election?
>
> Have you gotten a speeding ticket during the past twelve months?

MISTAKE 75: Stating a time frame that will be misunderstood by the subject.

EXAMPLE:

Last year, did you change your residence?

YES   NO

If the questionnaire was being filled out in December of 2002, and the move took place in January of 2001, does that count? 2001 was the last year, but maybe the examiner meant, "in the past twelve months."

EXAMPLE:

In the coming year, how likely are you to change your residence?

VERY LIKELY   SOMEWHAT LIKELY   NOT VERY LIKELY

If the questionnaire was being filled out in January of 2002, and the subject was intending to move in June of 2002, that would be the same year, not the next "coming year."

EXAMPLE:

Did you go to church last week?

If the questionnaire is being filled out on a Friday, and the subject went to church the previous Sunday, he might answer "no" thinking that Sunday was the first day of "this" week but he did not go to church on the previous Sunday (which would have been "last week."

SOLUTION: Always state time frames precisely.

During calendar year 2000, did you change your principal residence?

YES   NO

During the next twelve months, how likely are you to change your principal residence?

VERY LIKELY   SOMEWHAT LIKELY   NOT VERY LIKELY

Did you attend religious services during the past seven days?

YES   NO

MISTAKE 76: Using a time frame which obscures a confounding variable.

EXAMPLE:

Did you attend church or synagogue in the last seven days?

Suppose you conducted this survey over several weeks in April. You surveyed a Jewish sample during early April (before Passover). Later in the month you surveyed a Catholic sample (just after Easter). You might make the inference that Catholics attend religious services more often than do Jews.

SOLUTION: Schedule the distribution of your questionnaires in order to avoid any predictable peaks or valleys in the variable being measured. Another solution would be to anchor all the subjects to the same time period (e.g., the first week in April) but that will be more difficult for the later subjects, who will have less memory of what they did during that week.

MISTAKE 77: Assuming that there would be no gaps in the subject's time frame.

EXAMPLE:

When did you start working for the company?

The worker says 1991, and so the researcher assumes that the worker has spent over a decade with the company, but that worker might have quit and gone to work for some other employer at some point during that time, perhaps for the majority of those years, before returning to the company in question.

SOLUTION: If your goal is to measure total time with the firm, or in a specific occupation, measure that (and do so in such a way as to appreciate the skewed or truncated nature of the variable).

In total, how many years of service do you have with this company?

LESS THAN ONE

BETWEEN ONE AND FIVE

BETWEEN FIVE AND TEN

OVER TEN

MISTAKE 78: Asking for lifetime incidence when what you really want to know is recency, frequency or intensity.

EXAMPLE:

Have you ever stolen anything?

YES   NO

Have you ever been on the internet?

YES   NO

Have you ever felt sad, depressed or blue?

YES   NO

SOLUTION: Decide what is most relevant: recency, frequency, or intensity. Then ask a question that measures that.

In the past year, have you stolen anything worth more than $20?

YES   NO

Apart from email, how much time do you spend on the internet in an average week?

LESS THAN AN HOUR

ONE TO FIVE HOURS

FIVE TO TEN HOURS

OVER TEN HOURS

How sad, depressed, or blue are you feeling now?

EXTREMELY   VERY   SOMEWHAT   SLIGHTLY   NOT AT ALL

MISTAKE 79: Assuming too much knowledge of the topic on the part of the subject.

EXAMPLE:

How interested would you be in acquiring DSL?

The subject is thinking "What is DSL? What does it cost?"

SOLUTION: Provide enough information to so that the subject can make a realistic decision.

> DSL is a new technology for connecting your computer to the internet. It is about twice as fast as most modems, but not as fast as most T1 lines. DSL service would probably cost about $40 a month. How interested would you be acquiring DSL for your home?
>
> DEFINITELY INTERESTED
>
> POTENTIALLY INTERESTED
>
> NOT INTERESTED AT THIS TIME

MISTAKE 80: *Lack of a sufficient situational context for the subject to get a perspective on the question asked.*

EXAMPLE:

> Would you support the distribution of free condoms at your child's school?

That depends. Are you talking about where my 21 year old son goes to college or where my five year old daughter attends kindergarten?

EXAMPLE:

> Would you vote for Senator John McCain if he were running for President in 2004?

That depends. Who would he be running against? Are we talking about the primary or the general election? Are we assuming that he is the Republican nominee or that he is running as an independent?

EXAMPLE:

> Would you stay at a Motel 6?

That depends. Are there any other hotels around? Are they cheaper? Are they more comfortable? Am I traveling alone or with the entire family?

SOLUTION: Establish enough information so that the subject can put the question in the proper situational context.

Would you support the distribution of free condoms at your local high school?

In many cases, you will have to establish a hypothetical situation in order to set the proper context.

> Imagine that is the 2004 general election. The Republicans have nominated Senator McCain of Arizona, while the Democrats have nominated Senator Kerry of Massachusetts. Assume that there are no minor party candidates that you find attractive. For whom would you vote: Kerry or McCain?

> Imagine that you are on a long trip, driving for several days all by yourself. You now want to stop for the night, and there are only three hotels: Motel 6, one hotel which is a little cheaper (but you think might not be as nice), and one that costs a little more (but you think might be a little more comfortable). At which hotel would you stay?

One of the great advantages of establishing a hypothetical context is that it means that the questionnaire can be used to conduct an experiment. Two different hypothetical situations can be constructed (and this constitutes the independent variable manipulated).

> Imagine that you have just purchased a new computer, and then you remember that you need an anti-virus program. The clerk in the computer store has three brands: McAfee, Norton, and a brand X (which you have never heard of). The clerk says that they have a sale on brand X, making it $10 cheaper than either of its named competitors. Which brand would you purchase?

A separate groups design would have half of the questionnaires contain the above phrasing of the question, while the other half of the questionnaires would change one variable (e.g., the price might be $20 cheaper). If a significantly greater proportion of the subjects getting the $20 questionnaire would purchase brand X, then we know that the price drop will impact sales. If not, then there is no reason to further lower price.

A repeated measures experiment would accomplish the same task with one version of the questionnaire that included both questions.

Excessive precision

| MISTAKE 81: *Asking respondents for excessive precision.* |

EXAMPLE:

> How many times did you go to the grocery store last month?
>
> 0 1 2 3 4 5 6 7 8 9 10 11 12 13 14 15 16

Subjects who never went, or went once (when it was a memorable event) might remember, but just by using introspection, researchers should realize that most people would not know the answer to this unless shopping only and always occurred on a fixed schedule, such as Tuesday evening and Saturday morning.

SOLUTION: Use an ordinal scale.

> About how many times did you go to the grocery store in the last thirty days?
>> PROBABLY NEVER
>>
>> ONLY ONCE OR TWICE
>>
>> THREE TO TEN TIMES
>>
>> PROBABLY OVER TEN TIMES

| MISTAKE 82: *Asking subjects to create a composite assessment variable instead of global assessment.* |

| A composite variable is one quantifiable measure that is composed of two or more separate quantifiable measures. |

EXAMPLE:

> Give each of the clerical employees a score of zero to 100, considering their speed of data entry, but factoring in their error rate.

It is not clear how much weight each of these should receive, even assuming that the raters have access to objective and precise measures of speed and error rate. This approach to measuring performance becomes extremely problematic when different subjects are rated by different raters, and therefore, we cannot know if a given subject's rating says more about that subject's performance or that rater's perspective.

SOLUTION: If archival or field count data are available, the researcher should directly access those data and then calculate the composite variable using a consistent formula for all subjects. If raters must be used, we could get a global assessment of the rater's overall impression of the subject, and then the rater's evaluation of specific components. In this way, we could correlate to what extent overall ratings are associated with specific components.

Floor and ceiling effects

> *MISTAKE 83: Floor effect: the scale, question, or sample is constructed in such a way that most of the scores are at the low end of the response scale.*

EXAMPLE: A very difficult task was given to some new trainees. The performance measure was how many units could be assembled in ten minutes, but they had not yet been trained on that type of task. A majority of the sample was unable to correctly assemble even one unit, and giving them an extra ten minutes would not have helped.

SOLUTION: Having a pilot field test is important in that it can determine the proper level of task difficulty for the sample in question, and making sure that the phrasing of the question (and the responses provided) will not yield a median answer on either end of the scale.

> *MISTAKE 84: Ceiling effect: the scale, question, or sample is constructed in such a way that most of the scores are at the high end of the response scale.*

EXAMPLE: A very easy task was given to some veteran workers. The performance measure was how many of ten units could be assembled in an hour. A majority of the sample finished all ten units in less than half the time provided.

SOLUTION: Having a pilot field test is important in that it can determine the proper level of task difficulty for the sample in question, and making sure that the phrasing of the question (and the responses provided) will not yield a median answer on either end of the scale.

Changes from national poll

> **MISTAKE 85:** *Changing the phrasing of the question or answer from a national poll to make it more relevant to your sample. Once you change either the question or the answer format, you can no longer do a sample vs. norms comparison because you are then comparing "apples and oranges."*

EXAMPLE: One U.S. appliance manufacturer wanted to survey consumers about the perceived quality of one of its products. There was a national poll which had been done previously about the perceived quality of American manufactured products in general. The question was phrased something like this

Compared to what is offered by foreign competitors in general, how would you rate the quality of U.S. manufactured goods?

BETTER QUALITY   ABOUT THE SAME   POORER QUALITY

A third said "better" and the majority said "about the same."

This specific manufacturer decided that what it really wanted to know was the perceived quality of its product with respect to how it stood next to its chief competitor, a brand manufactured in Korea. So, the question was rephrased accordingly and the manufacturer was somewhat disappointed when the specific product did not fare as well against the Korean competitor.

SOLUTION: If a statistical comparison against the external norms is what is desired, then those norms can only be used if the phrasing remains the same. However, if the phrasing seems awkward or inappropriate, or unable to provide the information really desired, then go ahead and change the phrasing; but then realize that the statistical significance of the results must be tested by some other design: separate groups, repeated measures, or correlational, rather than sample vs. norms.

> **MISTAKE 86: Taking a question and answer format from an national poll using face-to-face or telephone interviews and administering it to the sample using written questionnaires.**
>
> Most national polls are based upon telephone interviews, and there are certain peculiarities of response that become more prevalent on written questionnaires.

EXAMPLE: A national poll of asked people about what they would be looking for in their next automobile.

What is the most important thing you will be looking for in your next automobile?

The results looked something like this

price	40%
safety	20%
comfort	15%
performance	15%
other factors	10%

One local auto dealer simply printed out the question and the above five possible answers and gave it to the people who walked into the show room.

Two problems were noticed. One was that a high percentage of the subjects had selected more than one factor. The national norms have percentages that add up to 100%, indicating that each subject selected one and only one factor. When an interview is being conducted over the phone, the actual marking of the data sheet is under the control of the researcher. If the subject mentions a couple of factors, the researcher can ask which of the two is more important, and then circle that one answer.

Another problem noticed was that the local auto dealer had a much higher of percentage of answers in the "other factors" category, which only left him wondering what those factors could have been. The reason why there was only a ten percent response rate on the telephone interview is that the researcher would read the question and then the first four possible answers. If the subject mentioned one of those, then that is what she circled. Only if the subject mentioned one of the factors not on the initial list was the "other" response circled. By printing out "other" as a possible answer, the written questionnaire encouraged the subjects to think about other factors, thus increasing that response category.

Political pollsters have noticed the same tendency. If you ask "Bush or Gore" you might get 45% for each with 10% volunteering the name of another candidate or "undecided" or "none of the above." If you put in those other names or possible answers, the percent for both Bush and Gore went down.

SOLUTION: The most obvious solution is to restrict the range of answers to those that the researcher most wants to deal with: a forced choice on the main items (avoiding the catch all "other factors" category). That may make the questionnaire more relevant to what the researcher wants to measure, but it makes it impossible to compare the sample's results with the external norms from the telephone polls. Therefore, if you are going to change the format from telephone to written, consider having some alternative design for testing hypotheses: separate groups, repeated measures, or correlational.

Changes from previous administration

> MISTAKE 87: *Changing the phrasing of a question from the first time it is administered to the next time it is administered makes it difficult to interpret the results of a repeated measures design.*

EXAMPLE: A 1992 student research project asked college student subjects how much time they spent each week on a computer. Now, another student wants to get contemporary data to compare to the older figures. However, she changed the question to

How much time do you spend each week on the internet?

Perhaps most of the time students spend on a computer they are connected to the internet, but not all. Perhaps they are writing term papers or playing games off line.

SOLUTION: A better approach would be to keep the original phrasing (if the goal is to compare to the previous study). Contemporary subjects can also be asked for how much time is spent on specific computer features (e.g., internet, writing, calculations). Then a different repeated measures design (comparing different features rather than different time periods) can be employed.

Loaded questions

> **MISTAKE 88: Loading a question so that it contains an argument that might influence the subject's response.**

EXAMPLE: Consider the question of whether or not public schools should use metal detectors similar to those used at airports and courthouses. Notice how the question can be written in one way to encourage agreement, and another way to encourage disagreement.

> Do you agree that schools should take reasonable security measures in order to protect your child in school from armed gang members?

> Do you think that school bureaucrats should be able to subject your children to the inconvenience and humiliation of making them pass through the kind of probing searches used on prisoners?

EXAMPLE: A tax cut was a major issue for Congress in early 2001. Notice how the question could be phrased to decrease or increase the support for the tax cut.

> Should hard working tax payers be given some relief from the crushing burden of supporting bloated federal bureaucracies?

> Should the rich be allowed to escape from paying their fair share of taxes, even if this jeopardizes social security and other vital programs?

SOLUTION: State the question in a clear and factual manner that enables the subject to comprehend what would be involved. If necessary, you can even include some arguments for, and against the proposal.

> Do you agree that schools should use security systems similar to those used at airports? Students and their backpacks would have to pass through metal detectors. The advantage would be that firearms and knives could be detected and stopped before they are brought onto the school grounds. The disadvantage would be the time and inconvenience of going through the lines, and then having a hand search if suspicious objects are identified.

> President Bush's package offers an across the board tax cut reducing the rates of all tax payers. Those who pay the most taxes would get the most savings. The Republicans claim that this will stimulate the economy. The Democrat package gives most of the tax savings to lower income tax payers. The Democrats claim that this is more fair. What should Congress do?

APPROVE THE BUSH TAX PLAN

APPROVE THE DEMOCRAT PLAN

DO NOT CUT TAXES. KEEP THE MONEY FOR GOVERNMENT NEEDS.

MISTAKE 89: *Using unbalanced responses in order to influence the subject's response.*

EXAMPLE: These examples use a balance scale to assess the proper level of funding. The wording of the first example implies that the amount is excessive, while the wording of the second example implies that the proposed amount is not large enough.

President Bush has proposed a $1.8 trillion tax cut.

How much of a tax cut should Congress approve?

THE ENTIRE $1.8 TRILLION AMOUNT

NO MORE THAN HALF A TRILLION

NO MORE THAN $100 BILLION

WE CANNOT AFFORD ANY AT THIS UNCERTAIN TIME

President Bush has proposed a $1.8 trillion tax cut.

How much of a tax cut should Congress approve?

AT LEAST $10 TRILLION

AT LEAST $5 TRILLION

AT LEAST $3 TRILLION

WE SHOULD SETTLE FOR THE MEAGER $1.8 TRILLION PROPOSED

SOLUTION: Try to put the status quo or the proposal to be evaluated in the middle. Also, make sure that the terms in the responses do not have subjective or prejudicial loadings.

President Bush has proposed a $1.8 trillion tax cut.

How much of a tax cut should Congress approve?

WE SHOULD HAVE A MUCH LARGER AMOUNT

THE $1.8 TRILLION IS ABOUT THE RIGHT AMOUNT

WE SHOULD HAVE A MUCH SMALLER AMOUNT (OR NONE)

False spectrum of one variable

> *MISTAKE 90: Assuming a nominal scale (A or B) defines a variable, when actually these are two separate variables, with a resulting answer pattern of A, B, both, or neither.*

EXAMPLE: I am often asked if I live in Mexico or California. The answer is both. I have homes in Acapulco and Toluca (Mexico) and Redlands (California).

EXAMPLE: I am often asked if the Bible contains historical fact that is literal, or stories which must be interpreted figuratively. Almost all denominations would agree that the answer is both. Where they disagree is which specific parts of the Bible are figurative, and what parts are factual.

EXAMPLE: I lived in Chicago for four years and was frequently asked if I was a Cubs fan or a White Sox fan. Perhaps most Chicagoans could be classified one or the other. I was neither (my loyalties sticking with my northern California origins, Go Giants!). I also liked the Oakland A's and wondered why Chicagoans could not also embrace both the National League Cubs and the American League White Sox.

EXAMPLE:

Do you attend high school or college?

In Redlands most 16 year olds are attending Redlands High School or East Valley High School while most of the 19 year olds are attending either the University of Redlands or Crafton Hills College. However, I see some young people in this age range who are not going to school anywhere at this point in their lives. I also have some 16 year old students attending evening courses at Crafton who are attending East Valley High School in the day.

EXAMPLE:

Have you been married or divorced?

Most of my students would answer "neither" (yet). Many are married and never (yet) divorced. Some are divorced (and were previously married). No one has been divorced who was not previously married.

EXAMPLE:

What kind of vehicle do you drive?

LARGE LUXURY CAR   SPORTS CAR   PICK UP TRUCK

Perhaps most drivers would fall into one, and only one of these categories. If someone only has a small economy car, that would not fit any of the above. I would have to answer "all of the above" because I have a Lincoln, a Mustang convertible, and a couple of pickup trucks.

EXAMPLE:

Do your children attend

PUBLIC SCHOOL   PRIVATE SCHOOL

Obviously, some people have no children, or their children are not yet of school age, or their children have already gone through school, or the children are being home schooled. Another possibility is that a parent may have one child in public school and another child in a private school.

SOLUTION: When categories are not mutually exclusive, use a separate question for each category. If the sample was composed of North Americans in Mexico, the question might be

Do you have a residence here in Mexico?

    YES, WE OWN OUR RESIDENCE

    YES, WE RENT A HOUSE OR APARTMENT

    NO, WE ARE JUST STAYING IN A HOTEL

`   NO, WE ARE JUST STAYING WITH FRIENDS

    NO, WE ARE JUST STAYING IN OUR R.V.

Do you have a residence back in the U.S.?

    YES, WE OWN A PLACE AND IT IS FOR OUR RESIDENCE

    YES, WE OWN A PLACE, BUT RENT IT OUT WHEN WE ARE DOWN HERE

YES, WE RENT A PLACE FOR US TO LIVE BACK IN THE U.S.

NO, WE DO NOT HAVE ANY PLACE BACK IN THE U.S.

In asking Chicagoans about baseball

Are you a Cubs fan?

FANATICALLY   LIKE 'EM   DON'T CARE   HATE 'EM

Are you a Sox fan?

FANATICALLY   LIKE 'EM   DON'T CARE   HATE 'EM

In asking about marital status

Have you been married?

YES   NO

Have you ever been divorced?

YES   NO

In asking about car ownership

Do you currently own (or regularly drive) a large luxury car?

YES   NO

Do you currently own (or regularly drive) a sports car?

YES   NO

Do you currently own (or regularly drive) a pick up truck?

YES   NO

For asking parents about the schools their children attend

How would you rate the local public schools?

EXCELLENT   GOOD   FAIR   POOR

Have any of your children ever attended a public school?

YES   NO

Do you currently have a child in a public school?

YES   NO

Have any of your children ever attended a private school?

YES   NO

Do you currently have a child in a private school?

YES   NO

EXAMPLE:

> One local school principal has decided to make available AIDS tests and condoms in her school. Would you favor that idea?

Maybe you support the idea of AIDS testing, but not the distribution of condoms (or vice versa). There are two separate issues here.

SOLUTION: Ask separate questions.

> One local school principal has decided to make available AIDS tests in her school. Would you favor that idea?
>
> DEFINITELY   PROBABLY   POSSIBLY   NO WAY

> One local school principal has decided to make available condoms in her school. Would you favor that idea?
>
> DEFINITELY   PROBABLY   POSSIBLY   NO WAY

It may turn out that the attitudes are directly correlated, and that subjects who support the distribution of condoms may be the same people who support AIDS testing, but that is a hypothesis to be advanced at the beginning of the research, and can only be confirmed by research which asks two separate questions.

MISTAKE 92: *Trying to determine an action, and the*
reason for that action, by using just one item.

EXAMPLE: After the problems with Firestone tires on Ford SUVs were reported, one polling organization wanted to find out if those problems made customers less likely to purchase Firestone tires.

> Are you less likely to purchase Firestone tires today because of the recent problems reported with Firestone tires on some Ford SUVs?

This is a special version of the "two variable in one item" problem. One variable is the subject's prediction of a future action (i.e., being less likely to purchase Firestone tires), and the second variable is the subject's own assessment of the reason for this change.

The real problem with this kind of question is that the results are easily misinterpreted. I would have answered "no" to the above question, and some Firestone executive might have falsely concluded that I was a loyal customer and that nothing else was necessary in order to keep my business. That inference would have been completely incorrect. I was not any less likely to purchase Firestone tires after the reports of problems, because I had already switched brands based upon some other problems I had previously encountered with Firestone tires.

SOLUTION: One possible solution is to employ a multiple nominal scaling for the response.

> Are you likely to purchase Firestone tires during the next twelve months?
>
> YES, I WILL PROBABLY PURCHASE TIRES, AND THEY WILL PROBABLY BE FIRESTONES.
>
> NO, I DON'T THINK THAT I SHALL BE PURCHASING ANY TIRES
>
> NO, I WILL PROBABLY PURCHASE SOME TIRES, BUT I AM CONCERNED ABOUT THE SAFETY OF FIRESTONES.
>
> NO, I WILL PROBABLY PURCHASE SOME TIRES, AND ALTHOUGH I THINK FIRESTONES ARE SAFE, I PREFER ANOTHER BRAND.

Obviously, this could get complicated in terms of different reasons and combinations of reasons.

A better solution is to break down the question into two separate issues: one measuring the likelihood of a future purchase, and the other exploring possible reasons for this.

> How likely are you to purchase Firestone tires during the next twelve months?
>
> VERY LIKELY   SOMEWHAT LIKELY   NOT VERY LIKELY
>
> Will you probably purchase a different brand of tires during the next twelve months?
>
> VERY LIKELY   SOMEWHAT LIKELY   NOT VERY LIKELY
>
> Compared to other brands of tires, how do you think Firestone tires rate in terms of their safety?
>
> ABOVE AVERAGE   ABOUT AVERAGE   BELOW AVERAGE
>
> Compared to other brands of tires, how do you think Firestone tires rate in terms of their handling?
>
> ABOVE AVERAGE   ABOUT AVERAGE   BELOW AVERAGE

136

Compared to other brands of tires, how do you think Firestone tires rate in terms of the comfort of the ride?

> ABOVE AVERAGE   ABOUT AVERAGE   BELOW AVERAGE

Compared to other brands of tires, how do you think Firestone tires rate in terms of their mileage: how far they can go before they are worn out and have to be replaced?

> ABOVE AVERAGE   ABOUT AVERAGE   BELOW AVERAGE

Compared to other brands of tires, how do you think Firestone tires rate in terms of their value: being a good buy for the price?

> ABOVE AVERAGE   ABOUT AVERAGE   BELOW AVERAGE

The first question measures the probability of purchasing the product. The next six items deal with reasons why. Someone who answers both of the first two questions "not very likely" is not in the market for tires. These subjects could be eliminated from the sample: they are not the target market.

The last five questions can be correlated with the first question to see what really is associated with the decision to purchase a Firestone tire (as opposed to some other brand). It may be that Firestone's main problem is that it is perceived as an overpriced brand (the last item).

The best solution whenever causation is being investigated is usually an experiment. Perhaps Firestone should be less concerned about the perceived problem, and more focused on the solution: what is it going to take to increase sales?

Would an improved tire design really improve sales, or has the name of Firestone be permanently tainted? If so, perhaps a name change is one aspect of the overall solution.

What kind of advertising and publicity campaign will it take to convince the public that Firestone has actually improved its tires? (or that its tires were falsely blamed for a problem which really should be blamed on Ford?)

> **MISTAKE 93: Assuming that subjects will identify the real reasons for their behavior.**
>
> This actually breaks down to two sub-assumptions. The first is that the subject is actually aware of the real reason behind something. The second is that the subject will be willing to honestly reveal the real reason on a questionnaire.

EXAMPLE: With many types of mental disorders, it does not help to ask the patient to self-diagnose. If I ask an Alzheimer's patient why he is having problems handling the tasks of daily life, he may blame his nagging wife. He may not understand, or accept, the idea that his brain cells are deteriorating. If I ask a schizophrenic why he is hallucinating, he may blame demons, the CIA, the Mafia, or Martians controlling his mind. If I ask a depressed woman the cause of her condition, she may blame an inattentive husband, but after four weeks on Prozac, he might look a lot better.

EXAMPLE: If a company has low morale among its workers, do not assume that a questionnaire will identify the cause (or a solution to improve morale). Workers in a low morale company, plant, department, or shift will find fault with everything, especially "the management" and pay. Comparable workers (same job, same pay levels) in a high morale environment will probably express more satisfaction with these factors, especially pay. Disgruntled workers think about how much they are paid: "They are not paying me enough to put up with all of this crap." Many non-profit organizations (e.g., parochial schools) have workers who accept low wages in order to have the opportunity for a fulfilling position.

SOLUTION: One solution is to use a correlational design: use one item to measure the condition (e.g., job satisfaction) and then have additional items in order to explore the possible causes.

What is your overall level of satisfaction with your job here?

VERY SATISFIED

SOMEHWAT SATISFIED

SOMEWHAT DISSATISFIED

VERY DISSATISFIED

What is your level of satisfaction with your pay?

VERY SATISFIED

SOMEHWAT SATISFIED

SOMEWHAT DISSATISFIED

VERY DISSATISFIED

What is your level of satisfaction with your fringe benefits?

VERY SATISFIED

SOMEHWAT SATISFIED

SOMEWHAT DISSATISFIED

VERY DISSATISFIED

What is your level of satisfaction with your supervision?

VERY SATISFIED

SOMEHWAT SATISFIED

SOMEWHAT DISSATISFIED

VERY DISSATISFIED

What is your level of satisfaction with your opportunity for advancement?

VERY SATISFIED

SOMEHWAT SATISFIED

SOMEWHAT DISSATISFIED

VERY DISSATISFIED

What is your level of satisfaction with the actual nature of the work you do?

VERY SATISFIED

SOMEHWAT SATISFIED

SOMEWHAT DISSATISFIED

VERY DISSATISFIED

If there is a causal connection (i.e., that dissatisfaction with pay causes overall dissatisfaction) we can expect that the correlation between these variables will be stronger. Unfortunately, we cannot always assume that a high correlation proves a causal relationship. In other words, if there is no correlation between job satisfaction and satisfaction with pay, then low pay is probably not causing problems with job satisfaction, but a high correlation between these variables does not mean that pay satisfaction leads to job satisfaction (it could be the other way around).

The best technique for inferring causal relationships is the experiment, but that could be most difficult (and expensive) when dealing with salary levels.

Chapter 8: response formats to avoid

Open ended

The vaguer the initial question, the broader the range of possible responses, including many which are well beyond the range of the variable the research is intending to measure. In the case of written or email questionnaires, the inappropriate response will not be recognized until after the subject is gone. The synchronous "real time" feature of face-to-face and telephone interviews would permit some immediate correction of inappropriate answers, but not without some embarrassment.

EXAMPLE: Race? This question can elicit responses such as "White" or "African American" as well as those such as "human" or "noble" or "100 meters."

EXAMPLE: Sex? This question can elicit responses such as "male" or "woman" as well as "heterosexual" or "not before marriage" or "only once a week" or "it depends on my mood."

EXAMPLE:

What fringe benefits would you like to see?

Subjects will respond in their own words, and these words will not necessarily form quantifiable categories. One might say "better health care" and another might say "coverage for dental" and then the researcher must decide if these two answers are similar enough to be lumped together in one category (and whether that category should also include "lower co-pays"). Does the response "more time off" go along with "vacation time" or "flextime"?

SOLUTION: The questionnaire should present the full range of acceptable responses as well as the question. If you want to determine whether a subject is male or female, pose those two items on the sheet of paper that the subject is filling out. When the goal is to quantify, and present results as percents, correlations, and medians, it is best for the researcher to present the subjects with specific responses.

What fringe benefits would you like to see?

    MORE MEDICAL COVERAGE

    EARLIER RETIREMENT

    LONGER VACATIONS

    FLEXTIME

    OTHER _____

The last question changes it into narrative data. Some subjects who might have circled one of the first four answers might now just mark other, thinking that it gives them a better opportunity to mention something more specific.

Vague

> **MISTAKE 95: *The use of scales which evaluate but do not suggest improvement.***

EXAMPLE:

An automobile dealer gave a written questionnaire to those who had taken a test drive in a small truck.

> What do you think about the trade off between fuel economy and performance?
>
> EXCELLENT   GOOD   FAIR   POOR

The truck was rated very low. What would it take to get the customers to purchase the truck? better mileage? a bigger engine with quicker acceleration?

EXAMPLE:

Faculty evaluations at the small college where I teach give a list of favorable characteristics of the professor, and the student has to fill out a bubble form indicating how well that professor has that characteristic. One of the items was "manages class time wisely."

One particular adjunct faculty member scored low on this item. We knew there was a problem, but the solution was not evident. Was she supposed to lecture more or have more discussion?

Another question dealt with whether or not the level at which the course was taught was appropriate. Students graded her poorly on that item as well. Does that mean she needs to come up a level or go down a level?

SOLUTION: Balance scales are appropriate when there is some golden mean that should be sought, and it is wise to avoid both extremes.

The automobile dealer could have used a question such as

All vehicles involve a trade-off in terms of fuel economy and performance (e.g., acceleration). How would you rate the XYZ model you just drove?

NEEDS BETTER	JUST ABOUT	NEEDS BETTER
FUEL ECONOMY	RIGHT	ACCELERATION

That information would be useful in redesigning the truck (if the overwhelming answer is one side of the other). If both extremes get a large number of votes, and few are in the middle, there may be a need to develop two separate models: a high performance version (perhaps targeted at younger males), and a fuel efficient version.

Let's go back to the faculty evaluation questions, which could be rephrased as follows.

What is your impression of the balance between lecture and discussion in this course?

TOO MUCH LECTURE	JUST ABOUT	TOO MUCH DISCUSSION
TOO LITTLE DISCUSSION	THE RIGHT MIX	TOO LITTLE LECTURE

What do you think about the general level of this course?

TOO ADVANCED	JUST ABOUT	TOO MUCH
FOR ME	THE RIGHT LEVEL	LIKE HIGH SCHOOL

The answer to this latter question might indicate a need to redesign the course, making it more challenging or easy. If both extreme answers receive high percentages compared to the middle, there might be a need for separate courses: one for majors and one for non-majors.

Checkmarks

> **MISTAKE 96: *The use of check marks.***
>
> Some questionnaires call upon subjects to select an answer by making a check mark. Other questionnaires are designed in such a way as to permit such check marks, or at least fail to suggest more appropriate alternatives. The bottom line is that the typical head and tail checkmark used by grade school teachers is extremely imprecise in indicating which printed answer is being selected.

EXAMPLE:

Check each of these words which describes you.

HONEST   STABLE   DETAILED   PRACTICAL

Suppose the subject puts a check mark halfway between "stable" and "detailed"? To which term does it refer? The one closer to the head or the tail of the check mark? Suppose the subject puts a "half" a check mark (just the tail) for some items, and a complete check mark for others, and an X for others. Should we give half credit for the half mark? Extra credit for the X?

SOLUTION: Instruct the subject to circle the desired responses. It is easier to tell which of several possible responses have been circled than check marked. I have never seen anyone use half a circle: an item is either circled or it is not.

Horizontal line

> **MISTAKE 97:** *Telling the subject to mark the point on a line which indicates the level of a certain variable.*

EXAMPLE:

> Here you see a horizontal line representing your satisfaction with your current internet service provider. Mark the point on the line which best describes your level of satisfaction.

LOW   HIGH

One problem is that if you do not put the words by the scale, subjects will get confused as to which end is good and which is bad. If you do put the words by the scale, up to a third of the subjects will circle (or check) the specific words rather than placing a mark on the line as instructed.

A greater problem is that subjects will not use the same types of marks. Some will use a dot that is hard to see. Some will use the (better) approach of an intersecting vertical line. Many will use a check mark which raises the question about what part of the check mark is to be used, the beginning of the head, the beginning of the tail or the end of the tail? A decision to consistently score is one way is not always that helpful, since the size of the check marks may vary greatly. (Indeed, some subjects may put a big check mark at the low end in order to indicate a lack of satisfaction, but a big check mark means a big tail that will extend over to the high end.)

SOLUTION: If you are intent on using the line approach, one way to improve this method would be to use a vertical line to measure the response. That is easier to grasp intuitively (top is good, bottom is bad). You can encourage that the mark be an intersecting horizontal line rather than a check mark.

A better approach is to replace the line with specific words on an ordinal scale.

How satisfied are you with your current internet service provider?

> EXTREMELY SATISFIED
>
> VERY SATISFIED
>
> GENERALLY SATISFIED
>
> SLIGHTLY SATISFIED
>
> SLIGHTLY DISSATISFIED
>
> GENERALLY DISSATISFIED
>
> VERY DISSATISFIED
>
> EXTREMELY DISSATISFIED

Another approach would be to show five different cartoon faces: big smile, little smile, flat expression, little frown, big frown.

MISTAKE 98: *Having subjects rank alternatives, 1st choice, 2nd choice, on down to last choice.*

The severity of this problem cannot be over-estimated. I have seen surveys in which the majority of the questionnaires had to be thrown out because the subjects had not filled them out in a way which could be used as scorable data testing the researcher's hypotheses.

EXAMPLE:

> Here are ten different makes and models of trucks. Rank them according to your priorities, best to worst.

Some people will confuse "rank" with "rate" and think that ten is the best and one the worst.

Even among subjects who correctly understand that a rank of 1 means first (best) and ten means last (worst), there are often misunderstandings about how to handle ties. Suppose I think that the best small trucks are Toyota and Ford, and see them as equal. I know that statistical procedures will require me to average ranks and give each a 1.5 ranking. Most subjects will just give both a 1, and then may give the next best truck a 2, etc. (with the result that the last placed truck will not get a rank of ten). Some subjects figure out that they better give out ten ranks, and just flip a coin to see if the 1 should go to Ford or Toyota.

SOLUTION: Have subjects rate each alternative on a numerical or ordinal scale, and then the researcher can later put the alternatives in appropriate ranked order for statistical tests such as the Spearman or Mann-Whitney.

How would you rate the Ford Ranger in its ability to meet your small truck needs?

MUCH BETTER THAN OTHER SMALL TRUCKS

SOMEWHAT BETTER THAN OTHER SMALL TRUCKS

CANNOT SAY THAT IT IS BETTER OR WORSE

SOMEWHAT WORSE THAN OTHER SMALL TRUCKS

MUCH WORSE THAN OTHER SMALL TRUCKS

How would you rate the Toyota Tacoma in its ability to meet your small truck needs?

MUCH BETTER THAN OTHER SMALL TRUCKS

SOMEWHAT BETTER THAN OTHER SMALL TRUCKS

CANNOT SAY THAT IT IS BETTER OR WORSE

SOMEWHAT WORSE THAN OTHER SMALL TRUCKS

MUCH WORSE THAN OTHER SMALL TRUCKS

Wrong end of scale

> **MISTAKE 99:** *Presenting levels on an ordinal scale in such a way that the subject will stop after fitting a lower level, even though a higher level would be more appropriate.*

EXAMPLE:

What is your highest level of educational attainment?

DID NOT COMPLETE HIGH SCHOOL OR G.E.D.

HIGH SCHOOL GRADUATE (OR HAVE A G.E.D.)

ATTENDED COLLEGE

EARNED A BACHELOR'S DEGREE

EARNED A GRADUATE OR PROFESSIONAL DEGREE (E.G., M.D., M.A., J.D.)

A subject who has earned a Ph.D. probably graduated from high school (and might circle that answer and stop), and certainly attended college (and might stop with that answer), and earned a bachelor's degree (and might stop there).

SOLUTION:

What is your highest level of educational attainment?

DID NOT COMPLETE HIGH SCHOOL OR G.E.D.

STOPPED AFTER GRADUATING FROM HIGH SCHOOL (OR GETTING G.E.D.)

SOME COLLEGE, BUT NO BACHELOR'S DEGREE YET

EARNED A BACHELOR'S DEGREE, BUT NO GRADUATE DEGREES

EARNED A GRADUATE OR PROFESSIONAL DEGREE (E.G., M.D., M.A., J.D.)

> **MISTAKE 100:** *Using terms that confuse the subject about which end of the scale is being measured.*

EXAMPLE:

Rate your internet service provider in terms of its quality of service (e.g., number of times you have difficulty logging on, getting dropped from connections).

HIGH   AVERAGE   LOW

Does the "high" mean higher than average quality? or higher than average number of problems?

SOLUTION: Use terms which clarify which end of the scale is good by indicating if it is quality or number of problems being measured.

Rate your internet service provider in terms of its quality of service (e.g., number of times you have difficulty logging on, getting dropped from connections).

EXCELLENT   GOOD   FAIR   POOR

Rate your internet service provider in terms of its quality of service (e.g., number of times you have difficulty logging on, getting dropped from connections).

BETTER THAN AVERAGE   ABOUT AVERAGE   WORSE THAN AVERAGE

Rate your internet service provider in terms of its quality of service (e.g., how often you have difficulty logging on, getting dropped from connections).

NEVER   RARELY   OCCASIONALLY   FREQUENTLY

MISTAKE 101: Asking for agreement on a mid-range position.

We will then know the position of subjects who agree, but we will not know the position of those subjects who disagree. Did they disagree because they see the mid-range position as too much or too little?

EXAMPLE:

During the late 1960s and early 1970s, polls consistently asked this question of U.S. voters.

Do you agree with the way that the President is handling the war in Viet Nam?

Over the course of the war, the percentage supporting the President consistently declined. What was clear, is that the American public was dissatisfied with the President's limited war policy of "graduated response." What was unclear is what the President needed to do to win that popular support. Should he become more of a "hawk" and more actively pursue military objectives, or should he become more of a "dove" and more actively pursue a de-escalation of the fighting in order to help the peace process.

EXAMPLE:

> Since the first moon landing, public support for funding of the space program has decreased.
>
> Do you agree with the federal government's current funding of the space program?

We know that those subjects who answered "yes" want to keep funding at current levels, but what do the increasing number of "no" answers mean? Do they want to drastically increase funding, or decrease funding?

SOLUTION: Use a balance scale with extremes on either end of the mid-range position.

> How do you think the President should handle the war in Viet Nam?
>
>   INCREASE MILITARY ACTIVITY TO WIN A VICTORY
>
>   MAINTAIN THE CURRENT LEVEL OF MILITARY ACTIVITY
>
>   DECREASE MILITARY ACTIVITY TO PROMOTE PEACE OR WITHDRAWAL
>
> Do you agree with the federal government's current funding of the space program?
>
>   NO, LESS MONEY SHOULD BE SPENT
>
>   YES, CURRENT LEVELS OF FUNDING ARE APPROPRIATE
>
>   NO, MORE MONEY NEEDS TO BE SPENT TO EXPAND THE SPACE PROGRAM

Sensitive questions

> **MISTAKE 102:** *Seeking a precise answer for sensitive questions.*
>
> Sensitive topics (e.g., personal hygiene, sexual practices, income) are more threatening to subjects when a precise (i.e., ratio scale) answer is sought. An additional problem with all ratio scales is that they assume that subjects are capable of a valid estimate.

EXAMPLE: Sexual behavior is a sensitive topic.

How many sexual partners have you had during your life?

EXAMPLE: Income is a sensitive topic among Americans.

What was your income last year?

SOLUTION: One solution is the use of an ordinal scale.

How many sexual partners have you had during your life?

NONE   ONLY ONE   2-4   6-10   11-20   MORE THAN 20

What is your estimate of net household income during 2001?

UNDER $20,000

$20,000 TO $39,999

$40,000 TO $59,999

$60,000 TO $79,999

$80,000 OR OVER

Overlapping categories

MISTAKE 103: *Overlapping categories on ordinal scales.*

EXAMPLE:

How old are you?

18-20   20-30   30-40   40-50   50-60   60

If I am fifty, do I circle 40-50 or 50-60?

SOLUTION: Have a clear cutoff for each level.

How old are you?

UNDER 20   20-29   30-39   40-49   50-59   60+

Inadequate categories

EXAMPLE:

What kind of automobile do you drive?

GENERAL MOTORS   FORD   CHRYSLER

"But, I drive a Nissan."

"But, I don't drive."

EXAMPLE:

For whom did you vote for in the 2000 election?

BUSH   GORE   NADER

"But, I voted for Buchanan."

"But, I did not vote."

SOLUTION: Put enough categories in to make the choice exhaustive, so that each and every subject will fit in one category. If necessary, put in one or more catch all categories.

What kind of automobile(s) do you drive?

GENERAL MOTORS   FORD   CHRYSLER   OTHER   NONE

Whom did you vote for in the 2000 election?

BUSH   GORE   NADER   OTHER   NO ONE

EXAMPLE: This is an inadequate range for measuring age in a community college student population.

How old are you?

18-20   21-24   25-29   30-39   40-49

Yes, we do have some students over 50, not very many, but a few. We also have some under 18.

SOLUTION: It is OK to have a questionnaire that has more extreme levels than that found in the sample (these categories will simply end up with a zero proportion).

How old are you?

UNDER 18   18-20   21-24   25-29   30-39   40-49   50+

Use open-ended categories at the low end and high end of the scale assures that no subject will be left out. It is possible that in a small sample, no one will end up selecting the extreme categories, in which case a zero proportion exists within that category. Of course, it is possible that any level in between can also end up with a zero proportion.

One limitation is that if there is an open-ended category (on the low end, high end, or both ends) the calculation of a mean level for the entire sample is inappropriate. The median and mode can still be used.

> ## MISTAKE 106: Using an overly broad "catch all" category with multiple nominal scaling.
>
> This may lump together very different subjects with different levels on the variable being measured.

EXAMPLE:

What kind of automobile do you drive?

GENERAL MOTORS   FORD   CHRYSLER   NONE OF THESE

The "none" category would include those who drive a Hyundai and those who drive a Rolls, and those who do not drive. This question is acceptable as written only if we don't care about any of the "none" respondents, and effectively want to exclude them from the final sample.

Who did you vote for in the 2000 election?

BUSH   GORE   NADER   NONE OF THESE

The "none" category would include those who voted to someone to the right of Bush and to the left of Nader, and lump them into one category, along with those who did not vote for anyone. This question is acceptable as written only if we don't care about the "none of these" respondents and want to exclude them from the final sample.

SOLUTION: The overly broad category is useful only if it is your intention to exclude all those subjects from data analysis, otherwise, it is important to lump them into useful groups that will be treated differently.

What kind of automobile do you drive?

GENERAL MOTORS   FORD   CHRYSLER   OTHER   NONE

Whom did you vote for in the 2000 election?

BUSH   GORE   NADER   OTHER   NO ONE

Improper use of numbers

> ## MISTAKE 107: Using numbers instead of categories.
>
> This is a great temptation, especially when the subject has been instructed to fill in bubbles, and the top of each row has a number.

EXAMPLE:

What is your religious affiliation?

1 = CATHOLIC   2 = PROTESTANT   3 = JEWISH   4 = OTHER

Even assuming that the four above categories are adequate and appropriate, the use of the numbers can create unnecessary confusion, not just for the subject filling out the questionnaire, but also for those interpreting the results.

SOLUTION: Wherever possible, stick with words when measuring categories or levels.

> ## MISTAKE 108: Using a numerical scale where 1 is high and 5 is low.
>
> Most people intuitively assume that a higher number is more, and therefore better.

EXAMPLE:

Faculty evaluations at the small college where I teach give a list of favorable characteristics of the professor, and the student has to fill out a bubble form indicating how well that professor has that characteristic (1 = excellent, 2 = good, 3 = fair, 4 = poor). I noticed that for most professors there would be inter-rater reliability. For example, Professor X was a great scholar in his field but a boring lecturer. Ninety percent of the students would agree that Professor X knew the content of his discipline (by giving him a 1 or 2 on that variable), but rated him more harshly on clarity of lectures (by giving him a 3 or 4 on that variable). Then I would come across some student evaluation forms which were the mirror image of the others: rating that same professor with a 3 or 4 on content, but a 1 or 2 on lecture clarity. An alternative interpretation of those responses is that the student got the scaling inverted. Additional evidence for the scale inversion came from the space on the form where comments were solicited from the students. "He may know what he is talking about, but we cannot follow him." When that comment appears with a 4 for content and a 1 for clarity, the student has obviously inverted the scaling.

SOLUTION: The easiest solution would be to use high numbers for better performance (or more agreement) and low numbers for lower performance (and poorer agreement). I also recommend using a zero to ten scale as opposed to a 1 to 5.

An even better solution is to get rid of the numbers entirely and replace with an alphabet or word-based format.

Give the instructor a grade for his/her knowledge of the content of this course.

A   A-   B+   B   B-   C   D   F

How would you rate the instructor's knowledge of course content?

EXCELLENT   GOOD   FAIR   POOR

> **MISTAKE 109: *Using a numbered response format and not being sufficiently clear about which response represents "not applicable.".***

EXAMPLE:

Faculty evaluations at the small college where I teach give a list of favorable characteristics of the professor, and the student has to fill out a bubble form indicating how well that professor has that characteristic (1 = excellent, 2 = good, 3 = fair, 4 = poor; 5 = not applicable). I noticed that for some students do not read all the response values once they see that 1 = excellent: they merely assume that the remaining numbers indicate poorer performance, and assume that 5 must indicate the worst possible performance. Some students will select 5 on some topics such as clarity of lectures (something which the course definitely involves) and then in the written comments section say something like "His lectures are impossible to follow."

EXAMPLE:

I constructed a personality test in which the subjects had to evaluate the personality of a famous historical figure on a zero to ten scale. One student who had chosen Abraham Lincoln rated him a zero on several traits, including "kind." I asked her if she thought Lincoln was so mean because he had chosen to fight the Civil War. Her answer was that she did not know Lincoln well enough to personally evaluate him as high or low on that trait, and so she had assumed that a zero meant not relevant, not understanding that she had scored Lincoln as being extremely UN-kind.

SOLUTION: In most cases, the best possible solution is to use words instead of numbers, and have the first answer that the subject sees be "not applicable."

How well organized are the instructor's lectures?

NOT APPLICABLE   EXCELLENT   GOOD   FAIR   POOR

In the case of our personality test, we decided to stick with the numerical format, and clarified in the instructions that in evaluating unknown traits, a score of 5 should be used because zero means extremely low. When that clarification of instructions was made, the number of 5s greatly increased, and 0s greatly decreased.

Columns

EXAMPLE: One marketing research company had a questionnaire that measured the consumers' perception about different brands of beer. The questionnaire was arranged in rows and columns. On the left was a short, descriptive phrase; on the top was the brand of beer: so each row was a different attribute, and each column was a different brand. There were at least a dozen attributes and at least eight brands of beer. There was no way that my screen could see all of the brands, or all of the attributes on one screen without scrolling.

	Miller	Budweiser	Coors	Corona
Refreshing	O	O	O	O
Expensive	O	O	O	O
Youthful	O	O	O	O

One risk is that the subject will assume that he is done, when he comes to the end, not realizing that he has to scroll down the page to see more attributes (rows).

Another risk is that the subject will assume that only those alternatives (columns) which appear on the screen are being measured, not realizing that he has to scroll right to see more possible examples of brands of beer.

The greatest risk is that the subject will scroll down to see the remaining attributes (or to the right to see the remaining alternatives) and then not remember what each row (or column) referred to, and end up clicking the wrong button.

SOLUTION: Put each item on one screen. Have that screen contain the instructions, the question, and all possible responses. The downside is that you have to make sure that there is rapid movement from one screen to another. Field test your design on several different screens (e.g., Mac vs. PC, Juno web with ad banners) in order to verify that everything is coming through on one screen.

> *MISTAKE 111: A questionnaire using a grid response format, and it is unclear if the subject is to select one row for each column, or one column for each row, or as many attributes which apply.*

EXAMPLE: One marketing research company had a questionnaire that measured the consumers' perception about different brands of beer. The questionnaire was arranged in rows and columns. On the left was a short, descriptive phrase; on the top was the brand of beer: so each row was a different attribute, and each column was a different brand. There were at least a dozen attributes and at least eight brands of beer. This questionnaire was administered online, and it was clear that the subject was supposed to click a circle describing a beer with an attribute.

What was unclear was whether the subject was suppose to select the one beer that best represented that attribute, or the one attribute which best described that beer. In this particular example, the researcher wanted the subject to select the one beer which best held that attribute. Because the questionnaire was administered online, I was able to figure out the answer through a process of trial and error. I said that Coors was "refreshing" but when I tried to select the same attribute for Corona, the checkmark for Coors disappeared. If this had been a paper and pencil questionnaire, I would not have received that immediate feedback, and the error would have only been discovered by the researcher after the questionnaire had been turned in (and would have to be discarded as unscorable).

SOLUTION: Avoid the grid scoring format. Each item should have its own clear phrasing of the question and specific alternatives.

If the goal is to find which one brand best expresses an attribute, this is the format which is best.

> Instructions: for each of the following items, you must circle one and only one answer. You might think that several brands might qualify (or that no brand fully qualifies), but you must indicate the one brand which best qualifies.
>
> Which one of the following brands of beer would you most associate with being "refreshing"?
>
> MILLER   BUDWEISER   COORS   CORONA

If the goal is to find which one attribute best describes a brand, then this is the format which is the best.

> Instructions: for each of the following items, you must circle one and only one answer. You might think that several words might describe a given beer (or that none of them fits perfectly) but you must select the one term which best describes that brand.
>
> Which one of these words do you most associate with Corona beer?
>
> REFRESHING   EXPENSIVE   YOUTHFUL   UNKNOWN

MISTAKE 112: *When using an adjective check list to build a score for a scale, putting all of the items for one scale together.*

EXAMPLE: We wanted to assess a job applicant's Holland Code. This is a six-dimensional measure of vocationally related interests (R,I,A,S,E,C). We came up with eight adjectives that measured each of the six dimensions. For example, the R factor (which is associated with hands-on jobs such as construction, farming, mechanics) would be indicated if the applicant circled adjectives such as stable, honest, practical, etc. The result was 48 adjectives arranged in six columns of eight adjectives.

R term #1	I term #1	A term #1
R term #2	I term #2	A term #2
R term #3	I term #3	A term #3
R term #4	I term #4	A term #4
R term #5	I term #5	A term #5

R term #6	I term #6	A term #6
R term #7	I term #7	A term #7
R term #8	I term #8	A term #8

We observed that in filling out a questionnaire, most subjects tended to start at the top and do the column on the left, and then the next, and finally the column on the right. What this resulted in was that the subject was deciding whether to circle the R factor items before moving on to the I factor, and so forth. In our particular example, most subjects tended to circle many more R items than C items, not necessarily because the person really had more of an R personality rather than a C personality, but because we had asked for 25 items to be circled, and subjects had used up almost all of their adjectives by the time they got to the last column.

Another problem is that if the adjective checklist is very long, the subject may grow fatigued or bored by the time the end of the long list is reached, resulting in greater carelessness (and lower validity of responses).

SOLUTION: The first solution we attempted to improve the validity of this check list approach was to reorganize the items. We now used four columns of twelve items each. Each factor was defined by rows, instead of columns (e.g., the first couple of rows contained the R terms as shown below).

R term1	R term2	R term3	R term4
R term5	R term6	R term7	R term8
I term1	I term2	I term3	I term4
I term5	I term6	I term7	I term8
A term1	A term2	A term3	A term4
A term5	A term6	A term7	A term8

This meant that each column would contain each of the six factors, and tended to distribute the overall number of responses between the factors more evenly.

Eventually, we administered this test by computer, each term appearing individually on a separate screen, giving the subject an interval level response format (zero to ten) for each. Then we added the scores of all the R terms to get a composite R score.) We maintained this order of the presentation of the items, (so that if the fatigue factor resulted in the subject giving one answer over and over again on the later items, this would not impact the relative levels of the composite R or C factors).

Number of desired responses

> **MISTAKE 113:** *Using an adjective check list without clarifying how many adjectives should be checked overall.*
>
> This is not so much of a problem if the purpose of the research is to compare subjects on a given item: those who checked the item vs. those who did not. This is more of a problem if the individual adjectives are going to be added up to form some form of composite scale score.

EXAMPLE: We wanted to assess a job applicant's Holland Code. This is a six-dimensional measure of vocationally related interests (R,I,A,S,E,C). We came up with eight adjectives which measured each of the six dimensions, for a total of 48 adjectives.

We observed that some job applicants would check very few terms, while others would check almost all of the terms. This frustrated our attempt to classify each person into his or high one main factor (e.g., an R type or an I type) since a person could be equally low (or equally high) on several factors. It also reduced our ability to compare subjects since a high checker would have more R points (even though R could have been one of his minor factors) and a low checker might have very few R points, even though it was his main factor).

SOLUTION: The first solution we came up with was that we told applicants to circle exactly 25 items. This meant that about half of the items would be checked, and the subject had to decide which of the 25 did the best job in describing himself/herself.

We found that about a fifth of the job applicants did not follow the instructions, and persisted in circling too many or too few adjectives. Some of these later admitted that they had not noticed the 25 limit in their rush to complete the test as quickly as possible. However, others admitted that they had seen the rule, but had chosen to ignore it because they thought it was more important to describe themselves in terms of only those adjectives which pertained (or in the case of high checkers, all of those adjectives which pertained). For our purposes, the fact that about a fifth of the applicants did not follow the rule was useful, since there were some positions in which following detailed instructions which one might not understand (or even disagree with) is an important aspect of the job.

A more mathematically elegant solution to this problem of high checking or low checking is to give each factor a weighted score proportional to the total number of check marks given. For example, a high checker might have selected forty adjectives, including four on the R factor scale, 10% of the total marks. A low checker might have selected the same four R factor adjectives, but out of a smaller total of sixteen marks made on the paper, so that would be an R factor of 50%.

Composite scoring

> **MISTAKE 114:** *Adding scores on individual items to create composite scores on a variable.*
>
> Some subjects will have a tendency to give low scores on most items, while some subjects will have a tendency to give high scores on most items

EXAMPLE: We wanted to assess a job applicant's Holland Code. This is a six-dimensional measure of vocationally related interests (R,I,A,S,E,C). We came up with eight adjectives which measured each of the six dimensions, for a total of 48 adjectives. Each adjective was then scored on a zero to ten scale (zero = not at all applicable; 10 = perfectly applicable).

Our hope was that most subjects would hover around the middle of the scale (about 5) on most of the items, with each subject using extremely high scores (9 or 10) for only a few items, and extremely low scores (0 or 1) for only a few items. While many people followed this approach in scoring, some subjects have a tendency to hover on the high end (giving out many 8's, 9's, and 10's), while others have a tendency to hover on the low end (giving out many 0's, 1's, and 2's).

SOLUTION: One solution which is frequently employed, but rarely works, is to tell the subjects that they must ration their total scores: e.g., only give out one 10 and one 0, only give out two 9's and two 1's, etc. If the total number of items are few, and if the scale is pretty narrow (say, 1 to 5) this might work with most of the subjects. The broader the range (0 to 10 is a broad range) and the greater the number of items, the more mathematically challenging will be the task of rationing the scores assigned by the subject.

The solution we employed was to allow the subjects to high end or low end their ratings, and then proportionately adjust their ratings. We wanted each scale (R,I,A,S,E,C) to have a potential range of 0 to 100, and a mid range of 50. We took the total number of points assigned to the eight R items (which could range anywhere from 0 x 8 to 10 x 8), multiplied times the ideal mid range of the composite score for all six scales (6 x 50), divided by the total number of scores assigned by the subject on all 48 adjectives. This is easily done with a computerized administration and scoring of the test.

Percent distribution

EXAMPLE:

Estimate the percentage of your grocery bill that goes for the following food items. (Do not consider non-food items purchased at a grocery store, or food bought in other types of stores.)

> CANDY, CAKES, PIES, OTHER DESERTS
>
> CHIPS, PRETZELS, OTHER DRY SNACKS
>
> FRESH DAIRY
>
> FRESH MEAT, POULTRY, SEA FOOD
>
> FRESH FRUIT AND VEGETABLES
>
> FROZEN FOODS
>
> CEREAL
>
> DRY FOODS (E.G., RICE, NOODLES, BEANS)
>
> SPICES AND CONDIMENTS
>
> FOOD IN CANS OR JARS

SOLUTION: Measure amounts of money or quantities purchased.

Estimate how much you spent during the last seven days for the following food items. (Do not consider non-food items purchased at a grocery store, or food bought in other types of stores.)

CANDY, CAKES, PIES, OTHER DESERTS

CHIPS, PRETZELS, OTHER DRY SNACKS

FRESH DAIRY

FRESH MEAT, POULTRY, SEA FOOD

FRESH FRUIT AND VEGETABLES

FROZEN FOODS

CEREAL

DRY FOODS (E.G., RICE, NOODLES, BEANS)

SPICES AND CONDIMENTS

FOOD IN CANS OR JARS

Another advantage to measuring exact quantities is that you can then compare individual categories between subjects, not just within subjects. The fact that I spend half of my food budget on fresh fruit and vegetables may make me sound like a major consumer, but that hides the fact that I spend little on them (and next to nothing on all the other categories). Someone with a larger household and greater total expenditures on all areas of food may have a smaller percent devoted to that category.

Chapter 9: suggestions on how to measure specific variables

Job

Many surveys relating to employment fail to distinguish between the different aspects: job title, job tenure, career preference, job satisfaction, department, performance, turnover, pay level, employer.

> **MISTAKE 116: Not clarifying that the question is about number of workers.**

EXAMPLE: A newspaper reporter was interviewing a supervisor of a government agency, and asked "How many people work here?" The answer was "about half."

EXAMPLE: When the same question was asked at another location, the answer was "Just ten in this office" but the reported wanted to find out how many were out in the field assigned to that office.

SOLUTION: Carefully distinguish the job aspect being measured.

For getting an estimate of the number of employees (as in above) the best phrasing would be

How many full-time, regular employees are there in your department?

      LESS THAN 20   21 - 50   51 - 100   OVER 100

How many full-time, regular employees are there at this factory?

      LESS THAN 50   51 - 100   101 - 300   OVER 300

How many full-time, regular employees are there in the ABC company?

      LESS 100   101 - 300   300-500   501 - 1000   OVER 1000

(Of course, a more precise figure might be obtained from the archives of the Human Resources department.)

> **MISTAKE 117: Confusing work and employment and unemployment.**

> Many governments define full-time employment as a job with regular scheduling currently amounting to at least 30 hours a week. People who do not have that level of employment, but are seeking it, are defined as unemployed. People who do not have that level of employment, but do not seek it, are classified as outside of the labor force (e.g., retired persons, students, homemakers, severely disabled, the institutionalized).

EXAMPLE:

Do you have a job?

My wife would answer yes, because she works hard preparing the meals, cleaning the house, and gardening. My father would answer no, because he retired from his regular job at IBM back in 1983. On the basis of their answers, my wife would be classified as employed, and my father would be classified as unemployed, but both of them should be classified as "not in the labor force."

SOLUTION: Use a multiple nominal scale.

What is your current employment status?

CURRENTLY EMPLOYED FULL TIME (NOT COUNTING WORK RELATED TO SCHOOL OR HOME)

NOT CURRENTLY EMPLOYED FULL TIME, BUT I AM PRESENTLY SEEKING FULL TIME EMPLOYMENT

NOT CURRENTLY EMPLOYED FULL TIME, AND I AM NOT SEEKING SUCH EMPLOYMENT

MISTAKE 118: Using inappropriate terminology for job title.

EXAMPLE: A study was made at a government agency attempting to determine the job satisfaction of clerical workers.

What is your job title?

CLERICAL   SUPERVISORY

Very few people circled clerical, and most circled neither of the options. In this particular agency, "clerk" referred to someone who provided window service to the public, and was not applied to data entry, filing, or secretarial positions. Indeed, the secretaries were referred to as "administrative assistants."

EXAMPLE: At another place of employment, the goal was to find out the job satisfaction of the janitorial and housekeeping staff.

What is your job title?

MAINTENANCE   OTHER

Very few circled maintenance. In that organization, the maintenance department were the mechanics and carpenters who went around fixing things. The people who kept things clean were known as the custodial department.

SOLUTION: This is a good example of where research should begin with a focus group or interview of key personnel in order to determine the proper titles.

The first example should have read

What is your job title?

DATA ENTRY   FILING   WINDOW CLERK   ADMIN ASSISTANT

For the second example,

What is your department?

CUSTODIAL   OTHER

MISTAKE 119: Using an inappropriate measure of job satisfaction?

EXAMPLE:

Do you like your job?

YES   NO

There are numerous problems with this question. The stem is unclear about which aspect of the job is being measured (probably overall satisfaction). Another problem is that the response format is binary nominal, and most workers think in ordinal terms on this variable. Another problem arises if the purpose of the research is to estimate future turnover (or labor management problems). A more direct question might be a more valid measure of those variables.

What is your overall level of satisfaction with your job here?

VERY SATISFIED

SOMEHWAT SATISFIED

SOMEWHAT DISSATISFIED

VERY DISSATISFIED

What is your level of satisfaction with your pay?

VERY SATISFIED

SOMEHWAT SATISFIED

SOMEWHAT DISSATISFIED

VERY DISSATISFIED

What is your level of satisfaction with your fringe benefits?

VERY SATISFIED

SOMEHWAT SATISFIED

SOMEWHAT DISSATISFIED

VERY DISSATISFIED

What is your level of satisfaction with your supervision?

VERY SATISFIED

SOMEHWAT SATISFIED

SOMEWHAT DISSATISFIED

VERY DISSATISFIED

What is your level of satisfaction with your opportunity for advancement?

VERY SATISFIED

SOMEHWAT SATISFIED

SOMEWHAT DISSATISFIED

VERY DISSATISFIED

What is your level of satisfaction with the actual nature of the work you do?

VERY SATISFIED

SOMEHWAT SATISFIED

SOMEWHAT DISSATISFIED

VERY DISSATISFIED

How likely is it that you will still be working with this company in twelve months?

VERY LIKELY

SOMEWHAT LIKELY

SOMEWHAT UNLIKELY

VERY UNLIKELY

Marital status

The important thing here is to clarify if we are looking for current marital status or past experiences.

EXAMPLE:

What is your current marital status?

SINGLE   MARRIED   DIVORCED

Putting the word "single" first runs the risk that someone who may use the term informally, might select it and then move on to the next question without even bothering to look at additional responses (e.g., divorced).

SOLUTION: This is a better response format for most adults.

What is your current marital status?

MARRIED   DIVORCED   WIDOWED   NEVER MARRIED

Keep in mind that the above format is a guide to current legal status. If your goal is to ascertain household status, this phrasing might be better.

What is your current marital status?

MARRIED, LIVING WITH SPOUSE

STILL LEGALLY MARRIED, BUT SEPARATED FROM SPOUSE

LEGALLY DIVORCED FROM SPOUSE

WIDOWED

NEVER MARRIED, BUT LIVING WITH LOVER

NEVER MARRIED, NOT LIVING WITH A LOVER

| MISTAKE 121: *Measuring current marital status when past experience of marriage of divorce is the variable to be measured.* |

180

EXAMPLE:

What is your current marital status?

MARRIED   DIVORCED   WIDOWED   NEVER MARRIED

Suppose a woman married at age 23, divorced her first husband at age 28, married her second husband at age 35, and was widowed from him at age 55.

SOLUTION: Especially with adults over age 40, a series of questions might be more appropriate.

Have you ever been married?

YES   NO

Have you ever been divorced?

YES   NO

Have you ever been widowed?

YES   NO

If the goal is to determine the number of times divorced, the question could be phrased.

How many times have you been divorced?

NEVER MARRIED, NEVER DIVORCED

MARRIED, BUT NEVER DIVORCED

DIVORCED ONCE, NEVER REMARRIED

DIVORCED ONCE, THEN REMARRIED

DIVORCED SEVERAL TIMES

One of the best ways to investigate these issues is by a private interview with flexible questioning.

How old were you when you first got married?

(if NEVER MARRIED, stop)

Are you still married to that person?

(if YES, stop)

(if the reason for the end of the marriage was not given, ask)

Was that marriage ended by divorce or death?

Did you remarry after that?

(if NO, stop)

When did you remarry?

(return to the above question about still being married)

Parenthood

> **MISTAKE 122:** *Confusing parental status with desire for children or living arrangements.*

EXAMPLE: One internet dating site asked a question about children and respondents had to check one of these four boxes.

> DESIRES CHILDREN
>
> HAS CHILDREN
>
> DOES NOT HAVE CHILDREN
>
> NOT AN ISSUE

Several different issues were conflated here, and it is possible that someone might want to circle more than one answer.

SOLUTION: Separate questions are necessary to measure separate variables. It is important to determine what we are really looking for. If the variable is the legal status of being a parent, the phrasing might work like this.

> Are you a parent?

> YES, I HAVE LEGAL RESPONSIBILITY FOR AT LEAST ONE DEPENDENT CHILD
>
> YES, BUT ALL MY CHILDREN ARE NOW ON THEIR OWN
>
> NO, I HAVE NO CHILDREN

If the goal is to determine living arrangements of the household, it might be phrased like this.

> How many minor children are in your household?
>
> 0   1   2   3   4   5+

If the goal is to determine an individual's desire to enter a marital relationship in which children will be added to the family (through birth or adoption), the question might be phrased.

Would you like to have (or adopt) children?

YES, AND I WOULD NOT MARRY SOMEONE UNLESS THEY WERE INTERESTED IN HAVING CHILDREN

YES, BUT I WOULD CONSIDER MARRYING SOMEONE NOT INTERESTED IN HAVING CHILDREN

NO, BUT I WOULD CONSIDER MARRYING SOMEONE INTERESTED IN HAVING CHILDREN

NO, AND I WOULD NOT MARRY SOMEONE INTERESTED IN HAVING CHILDREN

Age

> **MISTAKE 123:** *Measuring the variable of age with an inappropriate scale.*
>
> Age is frequently a relevant background variable of the subjects. It is useful to know the subjects age in order to verify that the sample is representative of the target population. It is also useful in assessing the impact of age on dependent variables such as attitudes and performance. However, the scaling used in the measurement of this variable must be appropriate to the sample and hypotheses.

EXAMPLE: Just giving a blank space for the subject to respond may yield a variety of vague and unanticipated answers:

"old enough to know better"

"still young enough to cut the mustard"

EXAMPLE: I have often done orally administered questionnaires with aged subjects in a nursing home, senior housing project, or senior recreation center. If there is some dementia present, the subject may get confused about current the exact age.

EXAMPLE:

How old are you?

    UNDER 20   20-29   30-39   40-49   50+

This question might be appropriate for measuring the age range of some samples of workers, consumers or voters, but not if the key cut off is 16 (for driving), or 18 (for voting), or 21 (for drinking), or 65 (for retirement). If a survey were designed to assess student attitudes about drinking, it would be important to identify who is under 21 and who has reached that milestone age.

SOLUTION: One solution is to ask for the year of birth. This is very effective in dealing with the problem of the elder who cannot recall age due to confusion about what year it is now. The problem with this approach is that some subjects may regard this as threatening anonymity.

Another solution is to ask for an range relevant to the hypotheses being investigated. For most student samples,

How old are you?

UNDER 18   18-24   OVER 25

because students most traditionally aged college students are between 18 and 24.

MISTAKE 124: Measuring the variable of age, when age is not really the variable you want to measure.

EXAMPLE: If you were distributing questionnaires in a mall, and asking age in order to infer if a student was attending high school, you might make an inappropriate inference. Most 17 year olds might be juniors or seniors in high school, but some have dropped out while others may have already graduated.

EXAMPLE: If you want to find out if students in California high schools were more satisfied with their educational experience in the 1980s than in the 1990s, don't just ask age and subtract to estimate when the subject was a student. Many current California residents went to high school somewhere else and then moved to California. Some of the current residents of California dropped out before finishing (or even starting) high school here.

EXAMPLE: If you were trying to see if patients on Medicare have less satisfaction with their health care than people on other health care plans, do not assume that people over 65 are on Medicare, and those under 65 are on other plans. The majority of people over 65 may be on Medicare, and the majority of people on Medicare may be over 65, but there are exceptions to both of these generalizations.

SOLUTION: Focus on the variable you really want to measure.

If you want to find out who is currently in high school, you must ask that question. (Remember that students might be out on summer vacation or "off track" waiting for the next term to begin.)

Are you currently enrolled in a high school?

YES, I HAVE SOME TIME TO GO BEFORE I COMPLETE IT.

NO, I HAVE ALREADY COMPLETED HIGH SCHOOL OR G.E.D.

NO, I DID NOT COMPLETE HIGH SCHOOL.

If you want to find out when some one attended a California high school, ask that.

Did you attend high school in California?

NO, I NEVER WENT TO HIGH SCHOOL.

NO, I ONLY WENT TO HIGH SCHOOL OUTSIDE OF CALIFORNIA.

YES, I GRADUATED IN THE YEAR OF _____

YES, I DID NOT GRADUATE, BUT LEFT IN THE YEAR OF _____

YES, AND I AM STILL ENROLLED IN HIGH SCHOOL

If you were trying to see if patients on Medicare have less satisfaction with their health care than people on other health care plans, ask one question about the level of satisfaction with health care and another question if the person is on Medicare.

In each of these examples, you might still want to ask a question about age. You might want to exclude from your sample people's whose age would be inconsistent with information about the other variables, such as someone who claims to be born in 1970 but claims to have graduated from high school in 1995.

Ethnicity

EXAMPLE: The term "race" may be objectionable to certain subjects. Indeed, some my write in the response "human."

EXAMPLE: Nationality can be understood to mean citizenship.

SOLUTION: Use a term such as "ethnicity" or "ancestry."

EXAMPLE:

Circle the one line which best describes your ethnic origin.

WHITE / CAUSASIAN / EUROPEAN-AMERICAN

ASIAN / FAR-EASTERN / ORIENTAL

BLACK / AFRICAN-AMERICAN

HISPANIC / LATINO / MEXICAN

NATIVE AMERICAN / INDIAN / ESKIMO

SOLUTION: The easiest approach is to transform the above question into a checklist by simply changing the instructions.

Circle each line which describes your ethnic origin.

WHITE / CAUSASIAN / EUROPEAN-AMERICAN

ASIAN / FAR-EASTERN / ORIENTAL

BLACK / AFRICAN-AMERICAN

HISPANIC / LATINO / MEXICAN

NATIVE AMERICAN / INDIAN / ESKIMO

Another approach would be to develop separate questions for each ethnic group. Each question could use a yes/no format or an ordinal scale.

How likely is it that you have some Black (African or African-American) ancestry.

CERTAIN   VERY LIKELY   SOMEWHAT LIKELY   VERY UNLIKELY

MISTAKE 127: *Using categories that are inappropriate for the sample or population.*

EXAMPLE: At most University of California campuses, the category "Asian" might be overly broad (and secure a plurality of the respondents).

SOLUTION: It might be better to break down that category into specific sub-categories (which could then be combined for comparison with national norms using the larger category.

Circle each line which describes your ethnic origin.

WHITE / CAUSASIAN / EUROPEAN-AMERICAN

BLACK / AFRICAN-AMERICAN

HISPANIC / LATINO / MEXICAN

NATIVE AMERICAN / INDIAN / ESKIMO

    CHINESE

JAPANESE

VIETNAMESE

CAMBODIAN

LAO

INDONESIAN

MALAYSIAN

EAST INDIAN

PAKISTANI

BANGLADESHI

OTHER ASIAN

FILIPINO

OTHER PACIFIC ISLANDER

Religion

> **MISTAKE 128: Assuming that religion is one variable.**
>
> There are several, separate, and measurable variables here: original affiliation, current affiliation, strength of commitment, doctrinal acceptance, frequency of activities.

EXAMPLE:

Do you approve of the Catholic religion?

> YES   NO

Even if this used an ordinal scale, such as a Likert, it is unclear if this refers to the doctrines of the Church, its emphasis on ritual, an approval rating of the Pope, etc.

SOLUTION: Break down questions about religion into specific aspects of affiliation, doctrine, and practice.

Here is a measure of affiliation.

"Are you, or have you even been, a member of the Catholic Church?"

> NEVER
>
> RAISED CATHOLIC, AND CURRENTLY PRACTICING
>
> RAISED CATHOLIC, BUT LEFT THE CHURCH
>
> RAISED CATHOLIC, BUT NOT CURRENTLY PRACTICING
>
> CONVERTED TO CATHOLIC, AND CURRENTLY PRACTICING
>
> CONVERTED TO CATHOLIC, BUT LEFT THE CHURCH
>
> CONVERTED TO CATHOLIC, BUT NOT CURRENTLY PRACTICING

Here is a measure of religiosity.

How important is religion in your life?

> VERY IMPORTANT   FAIRLY IMPORTANT   NOT VERY IMPORTANT

Here is a measure of activity.

Did you attend church, synagogue, temple, mosque, or other religious services within the past seven days?

YES   NO

Here is a measure of doctrinal acceptance.

What is the relevance of Jesus for you?

HE IS MY SAVIOR, SON OF GOD AND GOD THE SON

HE WAS GOD'S PROPHET, A HOLY MAN, BUT NOT GOD

HE IS A PERSONAL ROLE MODEL FOR ME, BUT JUST A MAN

HE WAS A TEACHER WHO HAD SOME GOOD IDEAS

ALTHOUGH HE BECAME A MAJOR FIGURE IN HISTORY, HE HAS NO RELEVANCE FOR ME PERSONALLY

I DO NOT THINK THAT HE EVER REALLY EXISTED

Education

> **MISTAKE 129:** *Measuring educational attainment by number of years of schooling, rather than degrees attained.*

EXAMPLE:

How many years of schooling do you have?

UNDER 8   8 TO 11   12   13   14   15   16   17+

"I have 9: one in kindergarten, four years in the first grade, and four years in the second."

"I only took fourteen years to get all the way to my masters."

"I have been taking classes at the local community college for over twenty years."

SOLUTION: Measure educational attainment in terms of degrees.

What is your highest level of educational attainment?

DID NOT COMPLETE HIGH SCHOOL OR G.E.D.

STOPPED AFTER GRADUATING FROM HIGH SCHOOL (OR GETTING G.E.D.)

SOME COLLEGE, BUT NO BACHELOR'S DEGREE YET

EARNED A BACHELOR'S DEGREE, BUT NO GRADUATE DEGREES

EARNED A GRADUATE OR PROFESSIONAL DEGREE (M.D., M.A., J.D., etc.)

Preference

| MISTAKE 130: *Confusing measures of performance with measures of preference.* |

EXAMPLE: A textbook publishing firm which was handling one of my previous textbooks took a survey which I had written for college professors and rewrote the response format. I had questions such as

How important is it that the textbook be a comprehensive list of recent material in the field?

How important is it that the textbook parallel your organization in covering the topics of the course?

How important is it that the textbook be readable and understandable by your students?

I suggested a response format such as

EXTREMELY   VERY   SOMEWHAT   SLIGHTLY   NOT AT ALL

My hypothesis was that the market was somewhat bifurcated, with professors in selective four year universities saying that comprehensiveness was the most important while community college instructors would say that student acceptability would be the most important factor.

Without my knowledge or approval, the marketing department of the publishing firm changed the response format before the questionnaire was sent out (possibly to save space). The answer was now

A   B   C   D   F

For college professors, these are familiar measures of performance, not preference. Seeing the response pattern, many instructors would have assumed that the question was asking something like

How well does the present textbook give you ... comprehensiveness?

instead of "Is comprehensiveness that important"?

SOLUTION: Use performance questions and responses when measuring performance, and preference questions and responses when measuring preference, and always distinguish between describing what is (performance) and what should be (preference).

Sexual orientation

> **MISTAKE 131: *Confusing sexual orientation with gender identity dysfunction.***
>
> Back in the 1960s, the American Psychological Association and the American Psychiatric Association decided that homosexuality would no longer be classified as a mental disorder. Homosexuality is a measure of the variable of sexual orientation (also known as sexual preference).

EXAMPLE: One researcher who wanted to identify a demographic and personality profile of individuals who might be potentially interested in sex change medical treatment asked this question.

> Are you satisfied with your heterosexual organs or are you gay?

Some heterosexuals are not satisfied with their bodies (somebody must be responding to those spam ads for penile enhancement). Most gay men and lesbian women are not interested in a transgendering sex change surgery.

SOLUTION: Directly focus on the variable to be measured: interest in the transgendering medical treatment.

> Suppose that your medical insurance covered the entire medical procedure (e.g., hormones, surgery) for changing your sex from male to female (or female to male). Would you seek this treatment?
>
> DEFINITELY   PROBABLY   POSSIBLY   NO WAY

> **MISTAKE 132: *Attempting to measure sexual orientation on a binary nominal scale.***

> Many individuals who have sexual experiences with members of the same sex consider themselves bisexual.

EXAMPLE:

What is your sexual orientation?

    HETEROSEXUAL   HOMOSEXUAL

SOLUTION: Use ordinal scaling.

What is your sexual orientation?

EXCLUSIVELY HETEROSEXUAL (ATTRACTED ONLY TO OPPOSITE SEX)

PREDOMINANTLY HETEROSEXUAL (MOSTLY ATTRACTED TO OPPOSITE SEX)

BISEXUAL (EQUALLY ATTRACTED TO MALES AND FEMALES)

PREDOMINANTLY HOMOSEXUAL (MOSTLY ATTRACTED TO SAME SEX)

EXCLUSIVELY HOMOSEXUAL (ATTRACTED ONLY TO SAME SEX)

> **MISTAKE 133: Measuring sexual orientation when the real goal is to measure sexual activity.**

EXAMPLE: The previous question is good for measuring orientation but not activity. It would be inappropriate to assume that someone who is exclusively homosexual has had many sex partners of the same gender.

SOLUTION: If the type and number of sexual contacts are the variables to be measured, clarify both the time frame and what constitutes a sexual contact.

A sexual contact is a person with whom you have had a relationship in which at least one of you touched (with any part of the body) the genitals of the other person. Thinking back over the past twelve months ...

How many sexual contacts have you had with different males?

    NONE   ONLY ONE   BETWEEN TWO AND FIVE   SIX OR MORE

How many sexual contacts have you had with different females?

    NONE   ONLY ONE   BETWEEN TWO AND FIVE   SIX OR MORE

Of course, if a different time frame is called for (e.g., lifetime behavior), then that needs to be specified in the question.

Section Four: Errors after the data come in

Chapter 10: coding errors

Aggregates only

> *MISTAKE 134: Recording the raw data as aggregates (totals of the entire sample) rather than subject by subject.*

EXAMPLE: One telemarketing manager hoped to find out if women were more likely than men to purchase the product over the phone. Sales agents were told to do the survey by keeping a record of the calls made.

Was the person answering the phone

    MALE   FEMALE

Did the call result in a sale?

    YES   NO

Several of the agents conducting the survey simply reported back these aggregate tabulations.

Total calls n = 100

Was the person answering the phone

    MALE = 25   FEMALE = 75

Did the call result in a sale?

    YES = 10   NO = 90

These data describe who answered the phone and how many sales were made, but the research question cannot be answered "Who was most likely to purchase: male or female?

SOLUTION: The best technique for data collection is to have one sheet of paper or card for each subject (in this case, each call which was answered). Notice that if this were a questionnaire, in which each subject filled out his or her own sheet, there would be one sheet of paper per subject. The raw data are then in a stack of papers. These papers can then be sorted by putting them into four smaller stacks to represent the four cells below.

Cross Tabulation of Variables

(subjects are people who answered phonecalls)

		Was a sale made by this call?		
		YES	NO	totals
G E N D E R	MALE	A 8	B 17	25
O F S U B J E C T	FEMALE	C 2	D 73	75
	Totals	10	90	N = 100

Now it is possible to see that most subjects stack up in cells A and D, so there is a correlation between calling a male and making the sale. Over a third of all the men who answered the phone bought the product, while less than 3% of the women were interested enough to buy it.

In the above example, it would be possible for the sales agents to directly enter the data into the coding table above, but it would become extremely difficult if we are going to collect more than two variables.

Another acceptable way to collect or code the variables is to enter the data directly into a spreadsheet program, such as Excel. Just remember that each subject should be a separate row, and each variable is a separate column.

Rounding up

EXAMPLE: One measure of productivity allowed for half units, but were then all rounded up to the next highest unit. When this is done consistently, and before aggregation, this introduces an upward error in the estimated aggregate totals.

SOLUTION: Here are the rounding rules for minimizing this problem. When rounding to the ones place, look at the tenths place. If the digit in the tenths place is a 0,1,2,3, or 4, round down (keep the digit currently in the ones place and ignore all the digits to the right). If the digit in the tenths place is a 6,7,8, or 9, round up to the next highest digit in the ones place. If the digit in the tenths place is a 5, and there is something to the right of the tenths place (other than pure zeros), round up. When there is nothing after the 5 in the tenths place, realize that you are exactly half way between the higher digit and the lower digit, and in order to avoid a consistent upward bias, always round to the even digit in the ones place. So, for example 6.5 would round down to 6 (because to round up to 7 would land on an odd digit), but at 9.5, round up to 10 (because to round down to 9 would land on an odd digit).

Digit to the right is a	So we should	Example
0,1,2,3, or 4	Round DOWN	3.1 becomes 3
6,7,8, or 9	Round UP	3.6 becomes 4
5 (and something follows the 5 in the next place)	Round UP	3.51 becomes 4
5 (and there are nothing but zeros to the right)	Round up or down to the even number	3.5 becomes 4 10.5 becomes 10

Elimination of subjects

EXAMPLE: A questionnaire was designed in order to determine the attitudes of young adults about sex. The questionnaire was administered to an evening class at a local community college, but five of the students in the class were high school students just taking that one class. Also, ten other students were over age 30.

SOLUTION: Begin by specifying the operational definition of "young adult" (e.g., between 18 and 24). Include an age related question on the questionnaire.

How old are you?

UNDER 18   18-24   25 OR OVER

EXAMPLE: A questionnaire was designed to measure personality traits associated with divorce. Two of the background questions were

Have you ever been married?

YES   NO

Have you ever been divorced?

YES   NO

One subject answered "no" to the first item but "yes" to the second, an impossible record. We cannot say if this error is due to the subject not carefully reading the questions or to willful lack of cooperation. In any event, this subject's data should not be included in the final tally.

EXAMPLE: A stop watch was used to measure the performance of fire fighters in training. It was not possible for the subjects to repeat the exercise. In one case, the observer forgot to start the watch when the subject began the drill. She then entered the score under time as 0. This datum is not a valid measure of the subject's performance, since he did not perform the task instantly. If this datum is kept in the survey, it will throw off the descriptive aggregate measures for the entire sample: mean, standard deviation, correlation with other variables.

SOLUTION: When a subject's measure on a key variable is clearly not valid, those data should be removed from the final tabulations.

MISTAKE 138: Eliminating subjects who run counter to the hypothesis you are trying to prove.

EXAMPLE: One teacher was trying to prove that bilingual education was effective. She found that of 33 students in her class, 20 did better and 13 did not. She then decided that eight of the 13 who did not improve, really were not trying, and therefore should not be included in the final sample. By excluding these eight cases, she was able to show a statistically significant improvement. This example is not just poor research methodology, it borders on being an unethical attempt to edit data in order to prove a point.

SOLUTION: Subjects should not be eliminated after the data collection unless there is reason to believe that measures on a background independent variable indicate that a subject should not have been included in the sample in the beginning, or that the measure of the dependent variable lacked validity and/or reliability.

MISTAKE 139: Eliminating subjects just to have two groups equal in size.

EXAMPLE: One manager was trying to see if there were differences between male and female members of his department. He obtained a total of 23 male questionnaires filled out, but only 8 female questionnaires filled out. He decided that he would make the two groups equal by throwing away 15 male questionnaires, leaving him with a sample size of 16: half male and half female.

SOLUTION: Nothing is gained in terms of the statistical significance or the representativeness of a sample by eliminating subjects. The two groups will be compared by measures of central tendency, such as percents or means, and that does not depend upon the two groups having an equal size. The larger each group, the more representative that group is of that portion of the population. The larger each group, the better the chance of attaining statistical significance.

Digitizing nominal and ordinal data

> *MISTAKE 140: Initially recording data on a binary nominal scale, and then hoping to convert it to a ratio scale.*
>
> Collected data can always be moved to a less precise scale. Ratio data can be moved to ordinal, ordinal to binary nominal, but we cannot go in the other direction.

EXAMPLE: One researcher asked if his customers were over or under age 30, and found two thirds were older. He then wondered how many were in their forties, but could not tell from these data.

SOLUTION: He should have started with a more precise scale:

How old are you?

     UNDER 20   20-29   30-39   40-49   50-59   60+

Those results could answer the more precise question of how many were in their 40s as well as the less precise question of how many were over 30.

Perils of partial data

> **MISTAKE 141:** *Scoring an unanswered item on a standardized test as a zero.*

EXAMPLE: The Geriatric Depression Scale has thirty yes/no items. Each item is scored one point if the response is depressive, and zero if it is not depressive. For example, "Are you generally satisfied with your life" gets one point for a "no" answer; "Do you have problems concentrating" gets one point for a "yes" answer. When the test is administered orally, the researcher can go through the questions one by one, repeating if necessary, in order to get recordable yes or no answer. When the subjects are simply given a paper form and a pencil with which to fill it out, a substantial minority will overlook (or choose not to answer) one or more items. (This percentage increases when subjects are physically or mentally deteriorated.)

If the subjects are just given a zero for each unanswered item, that means counting it as if it were non-depressed, whereas in fact it is the depressed patients who are somewhat more likely to leave items unanswered.

SOLUTION: The best solution is prevention. Having all of the answers in a visual pattern so that it is very easy to see that an item has not been answered is extremely useful. The lengths of the thirty questions on the GDS vary, and if the yes/no answers are put right after the questions, the result is a jagged pattern that is very hard to scan for completion errors. Putting all the yes/no answers in a column is readily scanable at a glance of the subject's own eye.

One solution I have heard advocated would be to give a middle value (.5 in this case) to any unanswered item. Another solution would be to simply delete those questionnaires that have any unanswered items, but this can become the majority in certain populations.

My preferred solution is highly effective, but it somewhat compromises anonymity: I look over the questionnaires before putting them in the ballot box. If there is an unanswered item, I say "You did not answer one. Look at this one right here" (putting my finger right where a yes or no should be. I try to avoid saying "You missed one" (which implies that they answered, but got it wrong) or "You forgot one" (which may highlight a deteriorating memory, and make them more depressed).

Chapter 11: statistical errors

Means

EXAMPLE: A questionnaire about religious affiliation gave Protestants a 1, Catholics a 2, and Jews a 3. The researcher ran a mean and summarized the sample as a 1.8, which assumes that the average of a Protestant and a Jew is a Catholic.

SOLUTION: The most appropriate measure of central tendency for nominal scales is the mode or percentages.

Jewish	5%
Catholic	30%
Protestant	65%

The proper presentation of percents as a measure of central tendency is 100 times part divided by whole.

EXAMPLE: The variable of income is highly skewed to the right. Imagine a sample of ten men walking out of a prayer group of a Baptist Church in an African-American neighborhood in Houston. The first nine men hold unskilled or blue collar jobs and make between 20 and 30 thousand dollars a year. The tenth man is George Foreman who may make close to 10 million a year. The mean of their incomes would add up all the incomes and divide by ten, giving the average man in this sample an income of over a million dollars a year.

EXAMPLE: Ten residents of Redmond Washington were asked their net worth. Nine had estimates which ranged from a hundred thousand dollars (if they had substantial equity in a home) to slightly in the negative range (if they had some credit card debt). The tenth person was Bill Gates, estimated to be worth about $50 billion. The made mean net worth about $5 billion. This is a problem known as a right skew: one high score in the sample (or group) makes the mean unrealistically high.

EXAMPLE: Of ten workers in a department, nine have productivity ratings very close to 90, but one had some serious last year and only scored a 40. That brought down the entire department's mean to 85. This is a problem known as a left skew: one low score in the sample (or group) makes the mean unrealistically low.

SOLUTION: The median is a measure of central tendency (the middle score of a sample or group) such that half of the scores are higher and half are lower. The median is not pulled toward the most extreme score. The mode (or modal category) is the most frequent score. It can be supplemented with percents. For example: 90% of the residents had net assets in the under $100,000 range, or half of the workers had productivity ratings above 90, and 40% were in the 80s.

Let me give some simple (but extreme sounding) advice on measuring central tendency. Never use the mean. Whenever there is a skew, the mean will be pulled in the direction of the extreme score. The mean is only appropriate when there is a standard bell curve distribution (where a few low scores are counter-balanced with a few high scores) and in that situation, the mean will be equal to the median. So, if you always use the median, you will not go wrong.

MISTAKE 144: Using means with truncated data.

> Means are only appropriate for data that conform to the standard bell curve (with two symmetrical tails). When tails are truncated, the mean is pulled away from the center of the distribution by the one remaining tail.

EXAMPLE: Suppose that job applicants who take a test of ability get scores that are distributed along a normal bell curve with a mean of 50 and a standard deviation of ten. However, the only people who are hired by the company are those who score above 40. The resulting distribution of scores among those applicants actually hired is therefore not normally distributed but has a right tail and a left truncation (produced by not hiring those who failed to score at least 40).

EXAMPLE: Suppose that another test that job applicants take is very easy, with a maximum score of a 100. Many applicants get a perfect score, and most are in the 90s. Since the maximum is so easy, this right truncation is known as a ceiling effect. (A test which is too hard, and most of the subjects score at the bottom of the range, producing a left truncation, is known as a floor effect).

SOLUTION: Median scores and percents in different ranges are much more descriptive of central tendency for truncated distributions.

MISTAKE 145: Using means for ordinal data.

EXAMPLE: Subjects expressed their attitudes on a five level Likert scale: strongly agree, mostly agree, neutral, mostly disagree, strongly disagree. When the researcher entered the data into a spreadsheet program that could only work with interval data, she scored each response as follows.

    5 = strongly agree
    4 = mostly agree
    3 = neutral
    2 = mostly disagree
    1 = strongly disagree

While this approach may be acceptable for quickly determining if there are any possible correlations, this is not an appropriate way of expressing central tendency. Expressing a mean of 2.8 may disguise a bimodal level distribution.

SOLUTION: Describe the answer in terms of a median level, or better yet, describe the percent at each level, such as

strongly agree	38%
mostly agree	12%
neutral	2%
mostly disagree	6%
strongly disagree	42%

Standard Deviations

> *MISTAKE 146: Calculating standard deviations or variances for variables measured on nominal, ordinal, or skewed scales.*

EXAMPLE: A questionnaire about customer satisfaction using a five point Likert scale reported a standard deviation. The meaning of that is hard to comprehend.

SOLUTION: The most appropriate measure of dispersion for nominal, ordinal, or skewed scales is percentages.

Miscalculating percents

> **MISTAKE 147:** *Reporting the results in terms of the actual number of cases instead of as a proportion of the total number of cases.*

EXAMPLE: One sexist fire chief presented data to show that women were physically unfit to become fire fighters.

> "Last year, only three women recruits could pass our rigorous tests of physical agility, strength, and speed.
>
> Compare that with the ten times as many men who passed."

This means nothing if ten times as many men also failed.

SOLUTION: In separate groups designs, there is usually an unequal number of subjects in each group. This was probably the case in the groups of male and female fire recruits who attempted the physical tests. What should be reported is the proportion of women who passed, versus the proportion of men who passed, and whether this difference between the two proportions was statistically significant (according to some inferential statistic such as the Fisher test, chi square, test of proportions, or Kolmogorov-Smirnov). The above data on fire fighter recruits could have been analyzed in a two-by-two contingency table

Cross Tabulation of Variables

(subjects are fire fighter recruits)

GENDER OF SUBJECT		Did recruit pass physical tests?		
		YES	NO	totals
	MALE	A 30	B 61	91
	FEMALE	C 3	D 6	9
	Totals	33	67	N = 100

About a third of men passed the test, and about a third of the women did, so there is no clearcut pattern of men or women proving superior on these tests.

> **MISTAKE 148:** *Expressing the difference between two central tendencies as "times greater" or "times more likely" instead of stating the proper measures of central tendency (e.g., part-whole percents, means, medians).*

> The popular media frequently do this, as do advocates of a particular cause determined to exaggerate the impact of these numbers on an unsophisticated public.

EXAMPLE: Another sexist fire chief looked for data to show that women were unfit to become fire fighters.

> "Last year, women recruits were shown to be twice as likely, compared to male recruits, to fail our tests of knowledge of fire science."

SOLUTION: The above way of describing the difference between the groups obscures how great the difference really was. The difference may have been great and statistically significant, for example, if the sample size was large, and if both groups were close to being equal in size, and if only a third of the men failed while two-thirds of the women failed.

However, the above presentation of data is usually used to exaggerate a small absolute difference. Suppose ten women tried out, and nine succeeded (a failure rate of 10%) while twenty men tried out, and nineteen succeeded (a failure rate of 5%): a difference that is too small to be statistically significant for such a sample size. Remember, a one percent failure rate is twice as high as a half a percent failure rate.

What should be reported is the proportion of women who passed, versus the proportion of men who passed, and whether this difference between the two proportions was statistically significant (according to some inferential statistic such as the Fisher test, chi square, test of proportions, or Kolmogorov-Smirnov).

> **MISTAKE 149: Reporting the results as a "percent more likely" in order to exaggerate the perceived impact.**

EXAMPLE: Go back to the previous example of the sexist fire chief with the same data.

> "Last year, women recruits were shown to be 100% more likely, compared to male recruits, to fail our tests of fire science knowledge."

This is an example of extreme distortion bordering on lying with statistics. The wording is consistent with percentage change, but conveys the notion of a part-whole percent: that 100% of all women attempting the physical tests failed.

SOLUTION: What should be reported is the proportion of women who passed, versus the proportion of men who passed, and whether this difference between the two proportions was statistically significant (according to some inferential statistic such as the Fisher test, chi square, test of proportions, or Kolmogorov-Smirnov).

MISTAKE 150: Reporting the results in terms of a difference between simple percents in order to minimize the perceived impact.

EXAMPLE: Training to reduce industrial accidents resulted in an accident rate that was only 4% below the previous rate. Should we conclude that the training was a failure? No, not if the previous rate was 5% and the new rate was 1% (and the sample size was large).

EXAMPLE: The number of Hispanics in the local community college is only 10% below their proportion in the surrounding community, and that does not sound like so much. It could be, if there are 20% Hispanics in the population, and only 10% in a large sample.

SOLUTION: Report central tendencies in the appropriate measures (e.g., percent, mean, median) and then give the significance of the difference between the two repeated measures, groups, or sample vs. norms.

MISTAKE 151: Presenting percents with an inappropriate base number as the whole.

EXAMPLE: Take the above example about the fire fighter recruits and the tests of physical ability and calculate the percents by having the numerator (part) being the number of female recruits who passed the test and the denominator (whole) being the total number of recruits (male and female together) who passed the test.

"Over ninety percent of the recruits passing our test were men."

This may be a significant finding, if the sample size was large and the population of recruits tested had an equal number of males and females. However, the design was probably separate groups, and there was probably a larger male group to start with. These results would mean nothing if ninety percent of those who failed also happened to be men.

SOLUTION: The proper presentation of percents in a separate group design is 100 times part divided by whole. The whole is the total number of subjects in the group (sub-sample). In the above case we are talking about a group of male recruits who took a test, and a group of female recruits who took a test. The part is the specific number within the group who have a specific characteristic (here, that means passing the test). This is the way that the above data should be reported.

Of the male recruits (n = 91) tested, 30 passed and 61 failed:

33% pass rate   67% failure rate

Of the female recruits (n = 9) tested, 3 passed and 6 failed:

33% pass rate   67% failure rate

Misinterpreting Correlations

> **MISTAKE 152:** *Ignoring or misinterpreting the sign in front of a correlation coefficient.*
>
> Correlation coefficients should be reported with either a positive (direct) or negative (inverse) sign in front of the decimal in order to indicate the direction of the relationship between the variables.

EXAMPLE: One researcher hypothesized a correlation between the workers' age and industrial accidents, thinking that older workers might be more at risk for industrial accidents (as they appear to be for automobile accidents). A correlation of -.23 (p < .05) was observed, so researcher could reject the null, and he then claimed that he had confirmed his hypothesis. Actually, the sign on the correlation coefficient shows that he had confirmed the opposite of his initial hypothesis: older workers were found to have fewer accidents than their younger counterparts.

EXAMPLE: One job shop offered a new form of computerized typing training using encouraging sounds and flashes after each few correctly typed sentences. The correlation between being in the group receiving this experimental encouragement and typing performance was -.35 (p < .05) so the job shop decided to spend more money in the training. However, the negative sign in front of the decimal shows that the relationship between the variables was inverse: those who had received the sounds and flashes actually did worse than the control group.

SOLUTION: Always put a + or a - sign in front of a correlation coefficient that you are reporting. Make sure that it matches the verbal report of the data. A positive correlation means the higher the subject was on the first variable, the more likely that subject was high on the second variable; and the lower a subject was on the first variable, the more likely that subject was to be low on the second variable. A negative correlation means the higher the subject was on the first variable, the more likely that subject was low on the second variable, and the lower a subject was on the first variable, the more likely that subject was to be high on the second variable.

MISTAKE 153: *Mislabeling the variables involved in a correlation.*

EXAMPLE: A researcher described the correlation (r = +.34, p < .05) between the variables of gender and performance as moderate, and concluded that men perform better on the job.

The researcher would be right if the correlation assumed that male gender was recorded as 1 and female gender as 0, but if female gender was recorded as 2, then the data indicate that the women have higher performance.

EXAMPLE: A research described the correlation (r = -.65, p < .01) between the variables of age and attendance as strong, and concluded that older workers are absent more often.

The researcher would be right if the correlation measured the dependent variable as attendance rate, but if the variable actually measured was number of days absent, that would indicate that the older workers had a better attendance rate.

SOLUTION: State correlations in terms of the actual measures of the variables.

> The correlation between male gender and performance was moderate (r = +.34, p < .05): men performed better than women.

> The correlation between worker age and absenteeism was high (r = -.65, p < .01): the older workers had better attendance.

Another approach which can add even greater clarity is to state the correlation in terms of percent differences or mean differences between two groups.

Men had a median of 86 units produced, while women only had a median of 76 units produced (p < .05).

Workers over age 60 were absent only 2% of the time, while workers in their 20s were absent 6% of the time (p < .01).

MISTAKE 154: *Assuming that a correlation is stronger than it is.*

Correlations represent the strength of a relationship between two variables (and the coefficients of correlation are usually represented by the letter r.) The closer the correlation is to zero, the weaker the relationship. The closer the correlation is to 1.00 (or -1.00), the stronger the relationship. The stronger the relationship, the fewer the exceptions to the trend; the weaker the relationship, the more exceptions to the trend.

EXAMPLE: One researcher found a statistically significant relationship that was excellent (p < .001). The correlation coefficient was r = .18. A large enough sample size can make a small correlation statistically significant, but that only means that the null hypothesis should be rejected (that the results cannot be explained by pure chance). The relatively weak correlation of .18 means that (although the trend could not be explained by random variation) many subjects were exceptions to the trend.

SOLUTION: How much of the variance of one variable can be explained by the other can be estimated by squaring the correlation coefficient, so a .70 correlation explains less than half (.49) of the variance. Here is a rough guideline for interpreting correlation coefficients.

Use the proper term for the strength of a correlation		
Range of coefficient	**Number of exceptions**	**Proper term for strength**
+1.00	None	Perfect
+.60 to +1.00	Few	Strong, high
+.20 to +.60	Some	Moderate
-.20 to 0.00	Many	Weak, low
0.00	So many, there is no trend	Zero, none
0.00 to -.20	Many	Weak, low
-.20 to -.60	Some	Moderate
-.60 to –1.00	Few	Strong, high
-1.00	None	Perfect

In physics, chemistry, biology, medicine, engineering, experimental psychology and clinical psychology, the expectation is that we should get high correlations. In studies of political science, consumer behavior, economics, sociology, anthropology, social and industrial psychology, it is hard to get high or even moderate correlations.

MISTAKE 155: Looking at a correlation coefficient without considering the slope and intercept of the regression line.

All the correlation coefficient reports is how strong the relationship is between the variables: it does not describe what that relationship is. A high correlation says that you can predict one variable from a knowledge of the other,

> but it takes a knowledge of the slope and intercept of the regression line in order to make a specific prediction.

EXAMPLE: One marketing researcher noticed a strong correlation ($r = +.84$, $p < .01$) between advertising expenditures on a product and the total revenue from that product. He urged the company to spend more on advertising.

The slope of the regression line was 1 (it went up at a forty-five degree angle): each additional dollar of advertising resulted in an additional dollar of revenue. If there was this one to one relationship between advertising and profit, the company would have only broken even on additional advertising expenditure, but this regression line was between advertising and revenues, and did not factor in the cost of producing and distributing the product. In other words, that regression line slope guaranteed that each additional dollar of advertising (when added to the other costs) guaranteed a loss on each additional unit sold.

SOLUTION: Plot the regression line (slope and intercept) so that specific predictions can be made.

> ### MISTAKE 156: Using the Pearson product moment correlation coefficient on inappropriate data.
>
> Most spreadsheets and statistical programs use the Pearson equation for correlation coefficients by default. However, this formula is technically only appropriate when both variables are measured in interval or ratio scales which are normally distributed. This turns out not to be a problem if both variables are measured in binary nominal scales (because the correlation coefficient calculated by the Pearson using 1's and 0's turns out to be the same as would be found by using a Phi coefficient appropriate for these data). It is also not that much of a problem for ordinal scales, if they also are normally

227

distributed (with most of the subjects stacking up in the middle of the range, and there are few outliers). It is even not too much of a problem if the sample size is large (which can mitigate the impact of skews and truncations). However, when the sample size is small, and the data are not close to being normally distributed, the use of a Pearson coefficient can not only exaggerate the strength of a correlation, it may even flip around the sign.

EXAMPLE: Consider these bivariate (X variable, Y variable) data for five subjects.

Subject	X score	Y score
Adams	4	6
Baker	3	2
Cisneros	5	5
Davis	6	7
Eng	1	10

Looking at the trend among the first four subjects, there is a slight trend for those who score low on one variable to score low on the other (a positive correlation), but the last subject is extremely in the opposite direction, so that the calculated Pearson coefficient is -.31.

SOLUTION: Especially when sample size is small (say, under 30) or when we have reason to believe that one or both variables suffer from a skew or truncation, it is better to use the Spearman rank order coefficient (which calculates with the above data to -.1). The good news is that this can be done easily with any spreadsheet or other program which uses the Pearson formula. Just enter each subject's rank instead of his score.

In this example, on the X variable, Davis is first with the highest X score, and would get the rank of 1. Cisneros would be in second place, Adams in third, Baker in fourth, and Eng would be last (have a rank of 5). On the Y variable, Eng is in first place (and gets a rank of 1), with Davis second, Adams third, Cisneros fourth, and Baker last (and gets a rank of 5).

Subject	X rank	Y rank
Adams	3	3
Baker	4	5
Cisneros	2	4
Davis	1	2
Eng	5	1

In general, if you always use the Spearman rank order coefficient, you cannot go wrong. When the variables are normally distributed, the Pearson and the Spearman tend to give coefficients of the same sign and with similar strengths. A serious distortion in either variable (e.g., skew, truncation) or one outlying case of bivariate data (as with subject Eng in the example above) can distort the strength (and sometimes even the direction) of the Pearson, but not the Spearman.

> *MISTAKE 157: Assuming that a correlation between two variables can only be linear.*
>
> In most cases, the relationship is non-linear, flattening out or accelerating at higher levels. In some cases, the relationship may be curvilinear (rising and then falling over the range of the predictor variable).

EXAMPLE: There is a direct correlation between income and life satisfaction: the more money people make, the more life satisfaction they tend to report. However, if we measure this correlation by the typical Pearson, linear regression formula, it would appear to be only a weak correlation. The reason for this is that the relationship between the variables is not linear. The first hundred thousand dollars brings a great deal of life satisfaction. The next hundred thousand dollars brings more life satisfaction, but not as much as the first hundred thousand brought: twice the income is not twice the life satisfaction. The next hundred thousand will bring another increment in overall life satisfaction, but not as much as the previous units of income. This corresponds to a basic principle of economics: the law of diminishing returns.

EXAMPLE: One industrial psychologist observed a moderate, direct correlation ($r = +.34$) that was fairly significant ($p < .05$) between a job applicant's previous experience in a sales position and his or her sales performance during the first year after being hired. If the above correlation is assumed to be linear, a job applicant with ten years of experience would be regarded as a much better bet than one with only five years of experience. The actual learning curves for most jobs are much shorter than that. A job applicant with one year of experience is a much better bet over one with no previous experience because the former probably learned a lot about the job (and himself) during that year. A job applicant with two years of experience might be a little better than one with only a year of experience, because she probably learned something additional in that second year on the job, but each additional year of experience will probably not as valuable as the previous year.

EXAMPLE: The Yerkes-Dodson law relates the variables of workplace stimulation and workplace performance. The law states that most workers achieve optimum performance at moderate levels of stimulation. At lower levels of stimulation, the workers get bored, and performance is impaired. At higher levels of stimulation, the workers get stressed out, and performance is impaired. This is known as a curvilinear relationship.

```
P                        optimum
E                         *   *
R                       *  at  *
F                     * midrange *
O                   *               *
R                 *                   *
M               *                       *
A     bore-   *                           *  stressed
N      dom  *                             *   out
C     * * *                               * * *
E
     L E V E L   O F   E N V I R O N M E N T A L
              S T I M U L A T I O N
```

SOLUTION: Plot out the scatter plot between the variables to see the shape of the relationship. Draw a scatterplot: each subject would be a bivariate data point with X coordinate being income, Y coordinate being life satisfaction. Because this is a positive (direct) relationship, we would see a rising curve, but it would level off as we moved from left to right because each additional unit of money brings a proportionately smaller increase in life satisfaction.

A better way to calculate correlation coefficients would be to transform one of the variables with exponents or logarithms before feeding the data into the Pearson equation.

Another approach would be to use a non-parametric measure of correlation, such as the Spearman rank-order formula.

One solution appropriate for the example about trying to predict on the job performance from the applicant's prior experience would be to give unequal stepwise increments for the predictor variable. In the case of job experience and job performance, it may be that a more valid predictor of on the job success would be to give three points for one year of experience, four points for two or more years of experience.

In the case of a curvilinear relationship, both a Pearson and Spearman might indicate that the correlation between the two variables was zero (i.e., these coefficients failed to note the trend). A superior approach here would be to recalibrate the variable of stimulation as three different

231

groups: low, moderate, and high levels of stimulation. Now calculate a mean (or median) level of performance for each of the three groups, and then use an Analysis of Variance (ANOVA) to show the significance of the difference between the means, or a Kruskall-Wallis test to show the significance of the difference between the medians.

> ## MISTAKE 158: Jumping to causal conclusions on the basis of correlational data.
>
> Many investigators will be too quick to infer that one variable caused the other just because the correlation was in the expected direction, and strong, and the data were statistically significant. All of these points are essential to inferring a causal relationship, but they are not adequate.

EXAMPLE: A newspaper reporter wrote this headline.

"Listen up bachelors, being single is hazardous to your health."

She then presented data from a large sample indicating that among men in their forties, the mortality rate was significantly higher among single men than among those who were married. What she had inferred was that being married was the cause, and living longer was the effect.

EXAMPLE: A market researcher for an automobile company noticed an interesting correlation among men: those men who have a pickup truck are more likely to have extensive sporting equipment (e.g., camping, hunting, fishing). What she could not figure out from the data is whether getting a truck caused them to buy sporting equipment, or whether having sporting equipment caused them to purchase a truck.

SOLUTION: Consider other possible explanations. Whenever two variables (e.g., X and Y) are correlated, there are four possible explanations. The first is that X caused Y. The second is that Y caused X. The third is that some other background variable (lets call it Z) caused both X and Y as collateral effects. (This explanation is known as a spurious correlation.) The fourth explanation is the null hypothesis (which means that any observed relationship reflects random variation, pure chance, and not a real underlying causal relationship between the variables).

We begin our causal analysis of a correlation by examining the last explanation, the null hypothesis through the use of inferential statistics. We estimate the probability (p) of random variation explaining the observed data. When the probability is less than one in twenty ($p < .05$) then we are fairly confident that the results are not explained by random variation, so we may reject the null hypothesis. Whenever $p > .05$, most social scientists would argue that we should accept the null hypothesis and not make claims about having found some cause and effect relationship.

In the above example, let us assume that the data came from a large sample, and that we have excellent significance ($p < .001$), so we must reject the null hypothesis as an explanation. But that does not allow us to reject the possibility of a spurious relationship between collateral effects. In other words, do the men really live longer because they are married, or are they married because they have certain characteristics that also tend to help them live longer?

Most men in their forties are married. Consider the kinds of men least likely to be married in their forties. These would be men who are incarcerated for criminal behavior, men who were divorced because they were drunks or had anger management problems, and gay men (who have higher rates of death due to AIDS and suicide). The happy, stable guy who does not run around to the bars each night is most likely to get and stay married, and most likely to live longer.

The best way to verify a cause and effect relationship between variables is to use an experiment (in which the independent variable is manipulated by the experimenter, and is not the result of the subject's own choice or behavior). This is easy to do with animal subjects, and in laboratory conditions. It would be impossible (and unethical) to do with the study of marriage and longevity: forcing half of the men to be married and the other half to be single just to see how it would impact the death rates. It would be possible (but expensive) to take a sample of men who do not yet have pickup trucks, give half of them trucks, and then observe if that group is more likely to go out and buy more sporting equipment.

Another technique is to do multivariate path analysis, and see which predictor variables have stronger correlations with both variables under study than the two have with each other. In the case of the pickups and sporting equipment, we might find that certain underlying personality traits predispose men to wanting a pickup truck and wanting sports equipment, or maybe the decisive factor is the demographic of geography: men in New York City are unlikely to have either pickups or extensive outdoor sporting equipment but men in rural Alaska are very likely to have a truck and sporting equipment.

Using inferential statistics incorrectly

> *MISTAKE 159: Using t-tests or Analysis of Variance when the scores are not normally distributed.*
>
> Both t-tests and ANOVAs attempt to infer if the difference between group means are statistically significant. If sample size is small and there are serious non-normalities (e.g., truncations, skews), the mean may not be the best measure of central tendency, and parametric measures of significance (such as t and ANOVA) may be inappropriate as well.

EXAMPLE: The sales of a small sample (n = 9) of real estate agents in one office was compared to the nationwide company norm. The one sample t-test found the data to be statistically significant ($p < .05$), but a non-parametric sign test was more cautious, urging acceptance of the null hypothesis.

SOLUTION: When the sample size is small, and the variable measured is distorted by a skew, truncation, bimodality, or other non-normality, use a non-parametric (distribution-free) significance test, such as sign, Mann-Whitney, Kruskall-Wallis, Kolmogorov-Smirnov, or Chi Square).

> *MISTAKE 160: Using chi square as an inferential statistic in samples that are small or have lop-sided grouping.*
>
> The chi square equations are, next to Pearsons, the most frequently used statistic, but they are only appropriate under certain circumstances.

EXAMPLE: Consider an example of fire service recruits passing a test of physical ability. The results were

Of the male recruits (n = 40) tested, 20 passed and 20 failed:

50% pass rate   50% failure rate

Of the female recruits (n = 4) tested, 3 passed and 1 failed:

75% pass rate   25% failure rate

Is 50% significantly different from 75%? A simple rows and columns chi square is considered inappropriate if the number of expected subjects in any cell is less than five. (For certain, if you have less than ten subjects in any of the groups, there will be at least one cell where the expected frequency is less than the five required).

SOLUTION: One solution is to employ the Yates corrected formula for the two-by-two chi square. A better solution is to employ the Fisher exact test for a two-by-two contingency table, and to use the Kolmogorov-Smirnov test for any other two-by-three (or more) contingency table.

One tail or two tail

MISTAKE 161: *Using a one-tail test when a two-tail is appropriate.*

EXAMPLE: After getting the p value for his data, John looked at a table column that was for .05 significance, one-tail. Since the z-score he had was 1.87, he noticed that he was larger than the critical value of 1.65, and so he concluded that he qualified for fair significance at the .05 level, and rejected the null hypothesis.

SOLUTION: Unless you have a situation that specifically qualifies for one-tail testing (e.g., a production control situation in which you do not care about overfilling a bag of produce, you are only concerned with guarding against under-filling), always use a two-tail test. If the only table you have access to is for one-tail use, simply double the p value shown. In the example above, John could have doubled the .05 at the top of the column, to realize that he only qualified at the .10 level of significance. To get to the .05 level for a two tail test, he should go to the one tail column for the .025 level and there John would find the correct critical value for z = 1.96. (Fortunately, when dealing with chi and F tables, you do not have to worry about tails.)

Size effect

> **MISTAKE 162:** *Assuming that a statistically significant relationship is the same thing as a major impact.*
>
> All that significance means is that the null hypothesis must be rejected (i.e., the data cannot be explained by random variation). It does not describe the strength of the relationship (correlation) between the variables. Statistical significance is more likely when a correlation is strong (or the differences between two groups is great), but it is also dependent upon sample size.

EXAMPLE: Sally was examining data from a large epidemiological study (n = 5,045). She found that the difference between the proportion of elders in their 80s with dementia and the proportion of elders in their 60s with dementia had an excellent significance (p < .001). Sally concluded that most people in their 80s have senile dementia, and that it must be caused by advanced age. Actually, most people over age 80 do not have dementia. The chronic brain syndromes leading to dementia (e.g., Alzheimer's disease) are not a direct result of aging, but they are more likely to occur in advanced age. The excellent significance reported by Sally was a function of large sample size.

SOLUTION: In addition to presenting the significance of the data (p value), also give a descriptive statistic which indicates "size effect" (such as Cohen's h, a correlation coefficient, or a difference between percents, means, or medians).

Trained workers had fewer errors compared to that of untrained workers (means of 2.3 and 4.3, respectively, $p < .01$).

Before the treatment, 68% reported a major problem, but after five weeks of treatment only 39% reported a major problem ($p < .05$).

The correlation between these two measures of job satisfaction was moderate ($r = .51$, $p < .001$).

Chapter 12: errors in interpretation, reporting, and presentation

Terminology and inference

> *MISTAKE 163: Using the term "bias" or "biased" to describe something inappropriate.*

EXAMPLE: John thought that some of the workers may have rated their manager poorly because he was Muslim, so he called that sample "biased."

EXAMPLE: Jim found that when he measured the number of accidents per worker in warehouse employees over the past five years, most had no accidents, about ten had only one, five had two, one had five accidents, and one had eleven. Jim called these data "biased" on the high end.

EXAMPLE: Jane analyzed the data of another researcher who had used a questionnaire in which some of the questions contained an argument which Jane thought might encourage subjects to give a high answer. Jane termed these items "biased."

Use the noun "bias" and the adjective "biased" only to describe samples that do not represent the desired population.

This sample of college students was taken during a night class, and has a bias toward older students (54% of this sample was over age 30, compared to only 27% of the student population at the university, $p < .01$).

Because there is such confusion about the meaning of these terms, for many audiences it would be better just to say "non-representative." John should have called those subjects "prejudiced" if they had rated their manager poorly solely because of his religious affiliation. Jim should have called the distribution of accident data "skewed right." Jane should have called those questions "loaded."

> **MISTAKE 164:** *The term "skew" is often misused to describe biased or inappropriate sampling, or loaded questions.*

EXAMPLE: Reporting that "the sample was skewed" because the subjects were drawn from the population of residents of a low income public housing project.

SOLUTION: Use terms such as "bias" or "non-representative" rather than skewed to describe problems in sampling.

> The sample was not representative (in terms of the background variable of income) because the subjects were drawn from the population of residents of a low income public housing project.

EXAMPLE: Reporting that "the results were biased because the subjects answered a question 'Do you support President Bush or Saddam Hussein'?"

SOLUTION: If the data have already come in, and the task is to report that the phrasing of the stem or response format might have distorted the results, that is a matter of the question being loaded.

Of course, a superior approach is to employ more neutral phrasing, such as

> How strongly do you support President Bush's Iraq policy?
> > STRONGLY SUPPORT BUSH'S POLICY
> > MODERATELY SUPPORT BUSH'S POLICY
> > NEUTRAL / DON'T KNOW
> > MODERATELY OPPOSE
> > STRONGLY OPPOSE BUSH'S POLICY

> **MISTAKE 165:** *Not recognizing the motive of researchers or subjects to exaggerate the results in order to conform to an organizational or political agenda.*

EXAMPLE: One principal who was trying to show that a new phonics oriented reading program had improved the reading ability of second graders asked teachers to rate the students' reading abilities at the beginning and end of the school year. One of the teachers returned the questionnaire, adding, "I said that everyone was 'excellent' after phonics, but if it would help any more, I can also rate everyone as 'poor' before phonics."

SOLUTION: If possible, it is best not to inform subjects of any stake that the researchers might have in the outcome of the results, though this may be quite difficult in terms of studies of the effectiveness of treatment interventions. Another solution is to utilize more objective measures of the dependent variable: perhaps valid and reliable objective measures of the pupils' reading ability. Another approach is to a separate groups design, with double blind so that neither the subjects nor the observers know which group got the new treatment and which group got the placebo.

> **MISTAKE 166: Use of incorrect pluralizations of terms associated with the reporting of research.**

EXAMPLE:

This data shows that we confirmed all of our hypothesises and our single main criteria.

SOLUTION: Use proper pluralizations. The above sentence should read

> These data show that we confirmed all of our hypotheses and our single main criterion.

In the field of information technology, data might be a singular word for "information," but within the social sciences, each bit of information is a datum, and the plural form is data. So, we never say "this data was" but "these data are." Criteria happens to be a plural word, with the singular form of criterion. Most words ending in "sis" have a pluralization of "ses" (e.g., hypothesis, analysis, crisis, basis, diagnosis, prognosis, neurosis, psychosis).

Singular noun	Plural noun	Adjectival form
Analysis	Analyses	Analytical
Bias	Biases	Biased
Crisis	Crises	Critical
Criterion	Criteria	Critical
Datum	Data	Empirical
Diagnosis	Diagnoses	Diagnostic
Hypothesis	Hypotheses	Hypothetical
Neurosis	Neuroses	Neurotic
Prognosis	Prognoses	Prognostic
Psychosis	Psychoses	Psychotic
Stimulus	Stimuli	Stimulating

MISTAKE 167: Using inappropriate terminology in describing the strength of a correlation.

EXAMPLE: Some students put all of their variables into an Excel spreadsheet. One pointed to a +.87, and explained to the other student that it was "good" and then pointed to a -.12 and said that it was "poor." The other student asked "What makes a correlation good: is it the .8 or the +"?

SOLUTION: Never refer to a correlation as good or bad, excellent or poor. The direction of a correlation can be described as positive (direct: both variables moving in the same direction) or negative (inverse: variables moving in opposite directions). The strength of a correlation is strong (indicated by a decimal close to +1.00 or -1.00) or weak (indicated by a decimal close to 0.00, whether on the positive or negative side).

MISTAKE 168: Using inappropriate terminology in describing the level of significance.

EXAMPLE: Some students put all of their variables into an Excel spreadsheet. One student asked if the data were significant. The other correctly pointed to the p value on the printout but incorrectly stated that she had a "high" level of confidence in the statistical significance.

The other student was confused: "How can your confidence be so high when the p number is so low, almost zero?"

SOLUTION: We should speak of significance only be using these terms: excellent (p < .001), good (p < .01), fair (p < .05), marginal (p < .10) and not significant (p > .10).

STATISTICAL SIGNIFICANCE (probability of the null hypothesis)

p = 1.00 - - - - - - - - - - (certainty)

p > .10   not significant   DO NOT REJECT THE NULL

p = .10 -

p < .10   marginal   DO NOT REJECT THE NULL

p = .05 -

p < .05   fair   REJECT NULL

p = .01 -

p < .01   good   REJECT NULL

p = .001 -

p < .001   excellent   REJECT NULL

p = 0.00 - - - - - - - - - - (impossibility)

> ## MISTAKE 169: *Assuming that a survey or experiment proves something although p > .10.*

EXAMPLE: Some students did a study on the impact of training on productivity: r = +.19, p > .10. One said, "The positive correlation shows that the more training, the greater the productivity, but it was only a weak correlation, so we just proved that it helped a little, but every little bit helps."

SOLUTION: When p > .05 (or .10 according to some standards) we accept the null hypothesis as our explanation. That means that we admit that we proved nothing, and that any observed correlation (or difference between means, or difference between percents) should be attributed to pure chance, random variation, not to the efficacy of any treatment being studied.

MISTAKE 170: *Assuming that statistical significance describes the degree of impact of the independent variable on the dependent variable.*

EXAMPLE: Some students did a study on the impact of training on productivity, and noticed that $p < .05$. One said "These data are significant. We just showed that in over 95% of the cases, training helped." The other student offered an even more exaggerated claim: "It means that the average worker increased productivity by over 95%. That training really works!"

SOLUTION: Realize that the p value only stands for the probability of the null hypothesis. When p gets below some predetermined level of significance (usually .05) that means that we must reject the null: we realize that the results are probably not due to pure chance, but admit that there is some sort of cause and effect relationship going on. However, the p value can be influenced by sample size (a larger size leads to a better level of significance) as well as by the impact of the independent variable on the dependent variable. The impact is best expressed by a correlation coefficient or a difference between measures of central tendency (such as means or percents).

MISTAKE 171: *Misunderstanding the concept of null hypothesis.*

EXAMPLE: One novice researcher was trying to show that male workers (compared to their female counterparts) performed better in a computer repair class. The statistician reported that the initial hypothesis could not be confirmed; that the null hypothesis would have to be accepted. The researcher wrote in his report

The null hypothesis is the opposite of my initial prediction, so I guess if we proved the null hypothesis, then that shows that we proved that the women did better than the men.

SOLUTION: Null means nothing. The null hypothesis is never proved, it is merely assumed (unless its probability is extremely small, in which case it is rejected). In other words, whenever we assume the null, we are declaring that we proved nothing.

Another way to understand statistical significance is that the probability that random variation (pure chance) can explain the resulting difference between the two measures (in this case, male vs. female performances). The only conclusion justified in the above example is that the difference between the men's and the women's outcomes was so slight that it could have easily been explained by random variation.

MISTAKE 172: Assuming that an observed trend is significant.

In order to be statistically significant, a trend must be very improbable. The less probable the outcome, the more statistically significant.

> if the probability were less than one in twenty ($p < .05$) the results have a fair significance
>
> if the probability were less than one in a hundred ($p < .01$) the results have a good significance
>
> if the probability were less than one in a thousand ($p < .001$) the results have an excellent significance

The most frequent error of novice researchers is known as Type I error (or alpha error): assuming that something is statistically significant when it is not (when $p > .05$). In other words, the most frequent mistake is to reject the null hypothesis without a sufficient basis for so doing.

EXAMPLE: A field count of customers ($n = 10$) walking into a computer store observed seven males and only three females. The researcher concluded that the store attracted a clientele of a predominantly male demographic.

EXAMPLE: A medical research team investigating a new anti-anxiety medication had ten patients on the medication, and another ten on an inert placebo. Eight of the medicated patients showed rapid improvement, compared to only four on the placebo. Some younger members of the research team cheered the results, exclaiming, "We have proved the new medication to be twice as effective"! When a more precise interval scale for the assessment of anxiety was used, it was found that the mean anxiety level of patients on the medication was 3.47 versus 7.34 for those on placebo. One of the team members then said "It really is twice as effective."

EXAMPLE: A company invested in some special training for its workers, and hypothesized that measured productivity would increase after training. Workers had a mean productivity of 78.5 before training, which increased to 79.1 after training.

EXAMPLE: A sales manager predicted that sales representatives (n = 21) who had a college degree would sell more than those sales representatives (n = 32) who had not completed a four year college degree. The college educated group had a mean monthly sales of $123,010 per rep, while the other group only had a mean monthly sales of $121,976 per rep.

SOLUTION: There is no way of telling if either of these differences are statistically significant, or whether we should attribute them to random variation (the null hypothesis) just by looking at the magnitude of the difference. But in each of the above cases, the higher measure is less than one percent greater than the lower measure, and that probably will not be a significant difference.

A type of statistical test known as an inferential statistic must be used in order to calculate or estimate the probability of the null hypothesis (of random variation accounting for the observed results).

In the above example about the customers, the probability can be calculated by the binomial distribution, sign test, test of proportions, chi square test, or Kolmogorov-Smirnov test. None of these tests estimate the probability of these results occurring by pure chance at anything less than .20. Therefore, the results should be reported as $p > .20$, and described as "not significant." The null hypothesis should be accepted.

Of course, the researchers might say, if we had looked at a hundred customers and found the same trend of 70% male, then that would be significant. Yes, but the point is that we cannot make that statement until we actually observe 100 customers. We cannot assume that the next ninety will have the same proportion as the first ten. We cannot make that assumption because we do not know if the first ten would be a representative same of the first hundred. The lack of statistical significance warns us that the trend in the first ten may be due to pure chance, just as if we had flipped a coin ten times and observed seven heads.

In the above example about the anti-anxiety medication, the same statistical tests could be used to estimate the significance of the difference between 80% improving and 40% improving. Even uncorrected versions of the chi square and two-sample test of proportions estimate p between .05 and .10 (and so not significant). The Yates' correction formula for chi square estimates a p between places $p = .17$. Similarly, we cannot assume that a sample of 100 patients would exhibit the same trend of 80% improvement. The interval scale for the assessment of anxiety may show a significant difference between these two groups of ten (depending upon how much intra-group variability there is in the individual scores). Whenever we use the more precise measures associated with interval and ratio scaling (e.g., means, standard deviations, Pearson coefficients) and the more powerful parametric significance tests (e.g., t, Analysis of Variance) we are more likely to commit Type I error.

> ## MISTAKE 173: Failure to acknowledge a significant trend.
>
> This is known as Type II error (or beta error) and is the practice of rigidly clinging to the null hypothesis and refusing to recognize significant data. Although it can be said that novice researchers rarely commit this error, there are times when it is so serious that we must be on guard against it.

EXAMPLE: A foreign medical research team investigating a new anti-psychotic medication decided to screen twenty patients (ten on the medication, ten on an inert placebo) for drug abnormalities. Four of the medicated patients showed such abnormalities, but only one of the placebo patients showed such an abnormality. The statisticians assured the research team that random variation could account for the difference between the two groups, and that the null hypothesis should be accepted.

SOLUTION: When we are dealing with human life, we should require a better margin of safety. It is true that the above study of twenty patients is inadequate to prove that the medication had a dangerous side effect, but it is also inadequate to prove that the medication is safe. The real limit to the study is the small sample size making it difficult to prove anything, and so the null hypothesis must be accepted (but the null hypothesis should not be seen as a guarantor of safety).

One possible solution, statistically, would be a larger sample size, but if this means exposing more human subjects to potential risks, that decision must be carefully weighted.

Another possible solution is to employ parametric inferential statistical tests (such as the t and ANOVA). These tests are more powerful (resistant to Type II error) but less robust (resistant to Type I error) compared to non-parametric alternatives (such as the Mann-Whitney, Kolmogorov-Smirnov, Kruskall-Wallis). In other words, the use of parametric inferential tests are more likely to give a better level of confidence.

If this sounds a little confusing about which test to use, run both a parametric test and a non-parametric test on the same data. If you have an adequate sample size and both tests agree, you have probably avoided both Type I and Type II error. If the tests disagree (and usually this means that the parametric says that the data are significant, while the non-parametric says not significant) then you have to decide whether you want to go with the parametric and run the risk of Type I error (rejecting the null prematurely) or Type II error (accepting the null and ignoring a real trend).

MISTAKE 174: *Ignoring the impact of intra-group variation.*

> Even if the differences between the groups or measures are great, the null might have to be accepted if there is a large variation within each group.

EXAMPLE: A company invested in some special training for its workers, and hypothesized that measured productivity would increase after training. Workers had a mean productivity of 78.5 before training, which increased to 88.7 after training. A closer examination of the findings revealed that most of the workers scored pretty much the same before and after, a few even decreased after training. Most of the big difference was produced by just a few workers who scored very low before, and made great gains after training.

EXAMPLE: A sales manager predicted that sales representatives (n = 21) who had a college degree would sell more than those sales representatives (n = 32) who had not completed a four year college degree. The college educated group had a mean monthly sales of $133,010 per rep, while the other group only had a mean monthly sales of $112,976 per rep. In both groups, some reps sold nothing during certain months, while other reps had an occasional half million dollar month.

SOLUTION: In the above examples, the higher measure is more than 10% greater than the lower. However, there is no minimum proportionate difference which can guarantee statistical significance. If there is a great deal of difference within each group or measure (as indicated by a large standard deviation, mean absolute deviation, variance, or range), the difference between the measures is not as significant.

The appropriate inferential statistical tests (e.g., t-test, analysis of variance) consider both the difference within each group (or measure) as well as the difference between groups (or measures). In general, the greater the variability within, the less likely that statistical significance will be obtained.

MISTAKE 175: Using the term "subject" to refer to the topic of the research.

EXAMPLE: Describing your research by saying "The subject of our research was car owner's attitudes about their new Chevy SUVs."

SOLUTION: Only use the term "subject" to describe the persons who are described by the data. (Many psychology journals now use the term "participants" instead of subjects.) In doing a questionnaire, the subjects are the persons who fill out the questionnaire, providing information about themselves: background information (independent variables) and attitudes (dependent variables). In the example above, the subjects of the research were the car owners who filled out the questionnaires. Their attitudes were the dependent variables being measured.

In laboratory experiments on rats or guinea pigs, the subjects would be the animals. Background factors such as gender, age, and species would usually be held constant. The independent variables manipulated would be the environmental stimuli or the chemicals injected. The dependent variable would be the animals' responses to these stimuli.

In field counts, the subjects are the persons whose behavior is being observed and recorded. If we are doing a field count of customers at a mall, the customers are the subjects. The independent variables would be the background factors (gender, estimated age, estimated ethnicity) while the dependent variables would be those things that are the result of the subjects' choices (e.g., what they are wearing, what stores they go into, whether they ask for assistance from a salesperson, what is purchased, whether or not a credit card is used).

In archival research, the subjects are the persons whose records contain background information or measures of preference or performance. The subjects of employment records would be workers. Background variables such as age, gender, or assigned training would be independent variables. The dependent variables would be productivity, absenteeism, accidents, supervisory evaluations. The subjects of a consumer database would be the individual customers. Demographic factors such as age, gender, and geography would be the independent variables. The kind of advertising campaign or price of the product would be independent variables. The decision to purchase on credit would be a dependent variable.

MISTAKE 176: Referring to a survey as an experiment.

EXAMPLE: One researcher measured the blood pressures of men and women in a laboratory, and called it an experiment.

SOLUTION: A survey measures variables. The gender of each subject (male or female) was measured. The blood pressure of each subject was measured on a ratio scale. An experiment manipulates the independent variable (which has the advantage of more readily determining its causal impact on the dependent variable). This manipulation can be in the form of random assignment (as in the case of separate groups experiments) or a before and after design (as in the case of repeated measures experiments).

charts, graphs and tables

EXAMPLE: John tried to use a pie chart to show the correlation between a consumer's income and the percentage of wealth invested in the stock market.

SOLUTION: Use pie charts only when data are distributed over a small number of categories (nominal scaling) or levels (ordinal scaling). A pie chart would apply to the above research only in that it could show the proportion of the sample that was invested in the stock market: the light gray has such an investment but the darker gray does not.

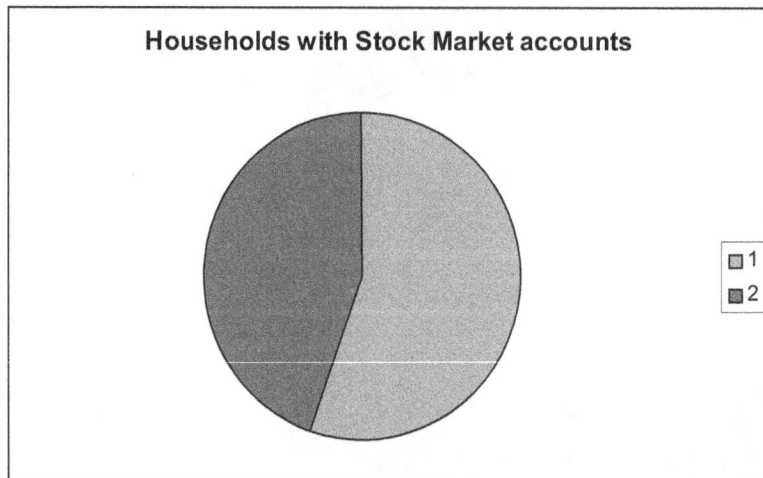

Use line graphs when showing how central tendencies (e.g., means, medians, percents) change over a time period.

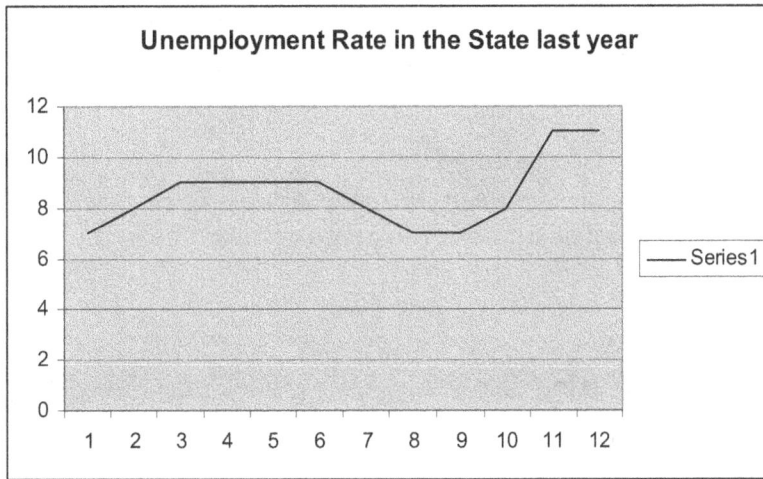

Unemployment Rate in the State last year

Use bar graphs (also known as column charts) to compare groups or repeated measures or sample vs. norms. Make each separate group a separate bar. Each separate repeated measure should be a separate bar. In a sample vs. norms design, have one bar for the sample and one for the norm. The height of the bar represents the aggregate or central tendency (e.g., mean, median, percent).

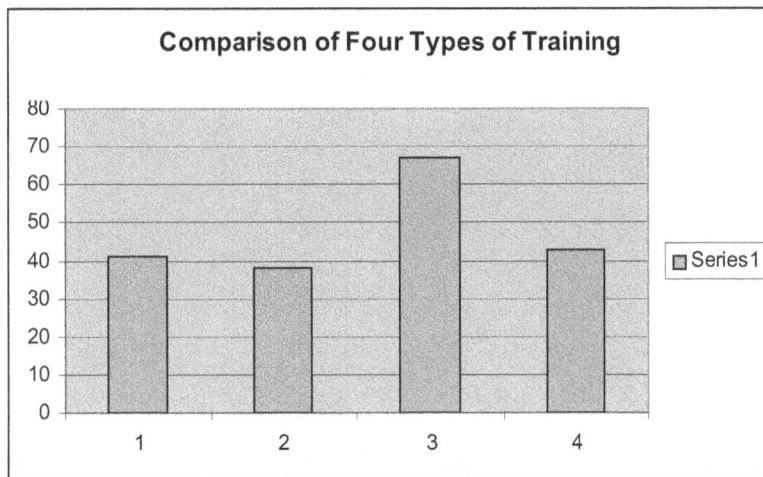

Comparison of Four Types of Training

Bar graphs are the most versatile. Anything a pie chart can do, a bar chart can do: just make each category a separate bar, and the height of the bar would represent the size of the piece of pie. Anything a line graph can do, a bar chart can do. Each time period would be a separate bar, and the height of the bar would be determined by the vertical point of the line.

Contingency tables are useful when we are trying to cross-tabulate data to show a correlation between two variables. These are manageable when we have a limited number of categories (or levels) for nominal (or ordinal) scaling.

Longitudinal Validation
(subjects are future drivers)

		ESTABLISHED MEASURE		
		GOOD driver	BAD driver	totals
P R I O R E X P E R I E N C E	Applicant has more than cutoff	A 20	B 5	25
	Applicant has less than cutoff	C 5	D 15	20
	Totals	25	20	N = 45

Scatterplots are useful for graphically depicting a bivariate correlation. One variable is given the X (horizontal) axis. This is usually the independent or predictor variable. The other variable is given the Y (vertical) axis. This is usually the dependent or criterion variable that we are trying to predict. It is also possible to correlate two dependent variables. Each subject is depicted as one data point, whose position is determined by the X and Y coordinates.

Worker performance

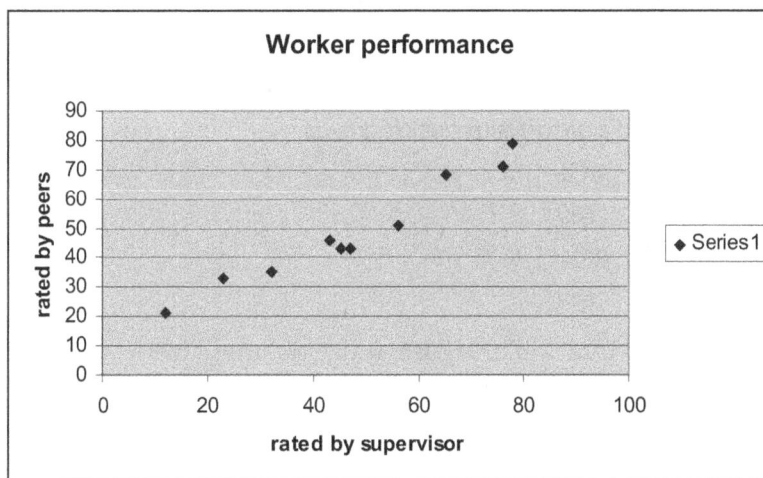

MISTAKE 178: *Setting up a table inconsistent with the research design. Designs are used to test a hypothesis by comparing separate groups, repeated measures, sample vs. population, or correlations.*

EXAMPLE: In reporting the results of the 2001 mayoral election between Hahn (the winner) and his chief opponent, Villaraigosa, one newspaper reported the outcomes like this.

Villaraigosa's supporters broke down as follows

White	48%
Latino	37%
Asian	5%
Black	7%

Did the candidate do better among Whites than Latinos because his policies appealed to a white agenda or because there were more white voters? We cannot tell from these data.

SOLUTION: In expressing percentages (which are a part/whole relationship), use the independent variable to determine the categories, and the dependent variable to tell us the size of each category. In the above example, the ethnicity of the voter is the independent variable, while the voter's decision to vote for Villaraigosa or Hahn is the dependent variable. Let us present the above data in this preferred format.

How Villaraigosa did among different ethnic groups

White	41%
Black	20%
Latino	82%
Asians	35%

Now it is obvious that Villaraigosa did extremely well among Latinos, but lost the other ethnic groups.

The write up

Good formal report writing is precise and concise. It generally follows a specific format and order: abstract (executive summary), introduction (review of the literature), hypotheses, method (subjects, apparatus, procedure), results, and discussion.

> **MISTAKE 179: Using a title that is cutesy, vague or rhetorical.**
>
> The purpose of the title is to clarify to the potential reader and indexing services exactly what the research covers.

EXAMPLES: A students who had conducted a survey on the death penalty thought about these possible titles.

> The case against the death penalty.
>
> Execution: is it right?
>
> Time to dance away from the electric chair?

SOLUTION: A good title describes the key dependent variable (in this case, the subjects ATTITUDES about the death penalty. Furthermore, it might also describe something about the sample or independent variables. Finally, it should use the terminology most commonly accepted by other researchers in the field. So, instead of "execution" or "death penalty" or "electric chair" the student authors of this report would find that "capital punishment" is used by most other researchers.

> Capital punishment attitudes of college students: the impact of respondent, victim, and perpetrator

> ## MISTAKE 180: An abstract that is a vague tease.
>
> Too many novice researchers take their cues from television newscasters who want to get viewers to stick around for the 11 o'clock news.

EXAMPLE:

What determines the attitudes of people about capital punishment? Is it whether they are male or female? Is it whether the victim was male or female? Is it whether the murderer was male or female? Keep reading our great report and you will find the answer to these questions.

SOLUTION: Since the abstract is the first thing that the reader reads (even if it is the last thing that the report writer writes), it should summarize the key findings. The reader will read the rest of the report if interested in how the results were obtained.

> ## MISTAKE 181: Using the abstract to summarize the literature review and discussion.

EXAMPLE:

Capital punishment (the death penalty) has been around along as formal law and punishment. Surveys about attitudes toward capital punishment are reviewed. The results of this study indicate the need for further research.

This abstract is just a summary of the introductory literature review and discussion sections. It tells us nothing about the unique contribution of this survey.

SOLUTION: Use the abstract to summarize the method and results sections. Give sample sizes (n's), descriptive statistics such as measures of central tendency (means, medians, percents), correlations (r's), and statistical significance (p's).

College students (n = 76) filled out a questionnaire involving different scenarios for murder, manipulating the variables of the gender of the murderer and the victim. Capital punishment for cases of single murder was supported by 48% of the entire sample. Support for capital punishment was slightly higher among male respondents (54% vs. 47%, $p < .10$), and for when the murderer was male (58% vs. 40%, $p < .05$). There was no interaction of effects between murderer gender and victim gender.

> ## MISTAKE 182: *Giving away the results in the introduction or statement of hypotheses.*
>
> The findings are reported in the results section, and summarized in the abstract.

EXAMPLE: Some report writers throw in a partial allusion to the findings when reviewing the literature or stating the hypotheses.

> Previous studies found that men were more likely to support the death penalty (which we our data, agreed with, as you shall see).

> Our first hypothesis was that men would be more likely to support the death penalty (and we proved that hypothesis).

SOLUTION: The review of the literature and initial statement of hypotheses should be written as if they were written prior to the actual conduct of research. Nothing describing the method or results of the present study belongs in these introductory areas.

> ## MISTAKE 183: *Inadequate description about the sample.*
>
> Since problems in biased sampling can create confounding variables, adequate description of the sample is essential.

EXAMPLE:

> We handed out the questionnaire to college students.

SOLUTION: Describe how the sample was selected, and give background information about the sample.

> College students (n = 76) exiting the library of a Midwestern public university were invited to participate in this study. The participants ended up having equal numbers of male and female (though this was not intentional on the part of

the researchers). Most of the participants were unmarried (72%), undergraduates (86%), and under the age of 25 (65%).

Note: Do not include information on the subject's attitudes about capital punishment in the description of the subjects. How they measured on the key dependent variable belongs in the results section, not the method section.

MISTAKE 184: Inadequate description about the questionnaire.

EXAMPLE:

Subjects filled out a questionnaire about capital punishment.

SOLUTION: Describe the questionnaire in great detail. If possible, include the instructions and major items on the questionnaire.

A quarter of the participants received the questionnaire below. A quarter received a questionnaire in which the man murdered a male coworker. A quarter received a questionnaire in which the murderer was a woman who murdered a man. Another quarter received a questionnaire in which the murderer was a woman who murdered another woman.

Instructions: this is an anonymous questionnaire, so do not write your name on this sheet. Please CIRCLE the best answer for each question.

What is your status here at the University?

UNDERGRADUATE STUDENT   GRADUATE STUDENT   OTHER

What is your gender?

MALE   FEMALE

How old are you?

UNDER 25   25 OR OVER

Are you currently married?

YES   NO

Suppose you were on a jury hearing a murder trial: a 40 year old man became jealous that female coworker was promoted over him, and poisoned her lunch. There was no doubt in your mind that he did it, so you voted to convict. Now it is the penalty phase of the trial. What do you vote for?

DEATH PENALTY   LIFE IMPRISONMENT

> ## MISTAKE 185: *Inadequate description of the procedure.*
>
> Readers should be able to understand how the researchers coded the data and calculated the statistics.

EXAMPLE:

> We took the questionnaires and crunched the numbers.

SOLUTION: Describe what was done in excruciating detail.

> The investigator stood outside of the library and approached subjects as they exited the university library during the late morning and early afternoon of a Wednesday during the fifth week of the semester. Each subject was asked if he or she was a student at the university. If so, the subject was told the nature of the survey, and invited to participate by filling out a sheet and depositing it the completed survey directly into the ballot box.

> A total of eighty questionnaires were filled out, but four had to be excluded from the final tabulation because of incomplete data or falling into the "other" category of student.

> Percentages supporting capital punishment were calculated for the entire sample, and also for groups determined by subject gender, murderer's gender, and victim's gender. Data were entered into two-by-two contingency tables. Since there was a minimum expected frequency of at least five cases per cell, statistical significance was inferred by the use of the Chi Square test for goodness of fit (with Yates' correction).

Note: the results of these statistics are not reported anywhere in the method section, but in the following section (results).

Index

New from Heuristic Books:

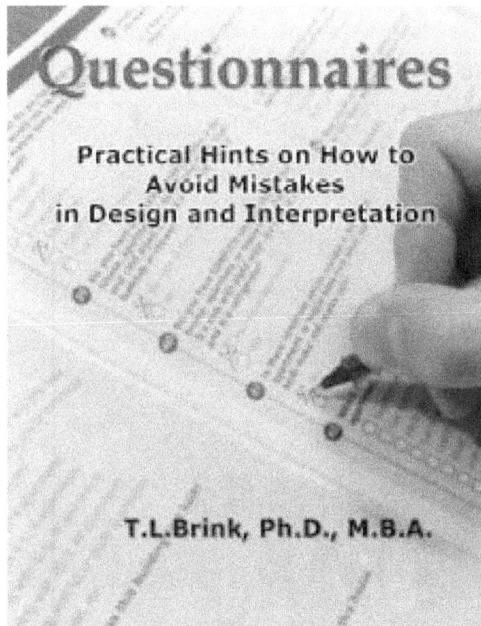

Questionnaires--Practical Hints on How to Avoid Mistakes in Design and Interpretation--By T.L.Brink, PhD, MBA (2004)

Buy this book before you implement that expensive survey!

Questionnaires is a practical guide to questionnaire design and interpretation with a problem-solution format. This book is different from others on this topic in that it is designed to be immediately helpful for people who have real-world constraints of budgets and deadlines. It focuses on eliminating costly errors in design rather than extensive theoretical aspects. The book is for people who have a need to design questionnaires: those who work in marketing research, public relations, opinion polling, and human resources. Questionnaires will be of practical use to executives in business, governmental, and not-for-profit organizations who have to make decisions based upon data from surveys. While it is not designed primarily for scholars and professors of research methodology its straightforward explanations should be immediately useful to students in graduate programs requiring a thesis for degrees in Business, Public Administration, Public Health, Psychology, and Sociology. T.L Brink holds a doctorate in psychology from The University of Chicago) and an M.B.A. from Santa Clara University. He is known worldwide for development of psychometric instruments for the assessment of psychopathology (e.g., the Geriatric Depression Scale), and has taught research methodology to psychology, sociology, business, public administration, and public health students at U.S., Mexican and Spanish universities. Brink is the author of several hundred articles and reviews in journals such as American Journal of Psychiatry, Academy of Management Review, Contemporary Psychology. He has several dozen encyclopedia articles on topics such as qualitative methods, management theory, measurement of religiosity, and has authored, co-authored, or edited nine previous books.

Coming soon:

Centers for Disease Control & Prevention

Youth Risk Behavior Survey
with Student Guide for Statistical Analysis in EXCEL

from Surveillance Summaries   June 28, 2002 / Vol. 51 / No. SS-4

Student Guide by
Robert J Banis, PhD, CMA

CENTERS FOR DISEASE CONTROL AND PREVENTION
SAFER · HEALTHIER · PEOPLE

Free video tutorials on Statistics techniques in EXCEL: These video tutorials are free to download, reproduce and distribute for nonprofit educational purposes. An adjunct resource for Statistics books from Heuristic Books at heuristicnooks.com. Watch for updates and more tools.

Winning with Statistics

Peter Drucker said, **"Disagreement is only possible when you don't have the facts. Get the facts."**

Statistics is a discipline for people who want to:

1. Think for themselves
2. Be objective in discovering the facts that may be hidden in a mass of data.
3. Argue persuasively for what they have learned.

To realize these benefits, the student of statistics has to be able to use and understand the tools. The tools provide the ability to discover relationships that may not be obvious in a quagmire of data, to use the rules of evidence and logic to test assumptions, to make a convincing case that assumptions are "false beyond a shadow of a doubt" —or, at least to ascertain how much doubt is warranted. There is no need to just accept other people's opinions that you hear on the news.

Discover for yourself—with step-by-step instructions, lots of screen captures and videotutorials at the author's companion website at the University of Missouri—St Louis— all the things you ever wondered about risky behavior by highschool students.

The tools are commonly available — as in Microsoft EXCEL — to become an expert at learning.

This book is more than just a guide to the Youth Risk Behavior Surveillance Survey (YRBSS) — one of the many sites on the web with Government data on current social issues. It's the beginning of an introduction to the world of discovering new knowledge. The tools provided using the YRBSS as a first example will apply equally well to the myriad new resources available to the modern student on the web.

+ Is smoking marijuana related to depression and suicidal tendencies?
+ Are males or females more violent towards their companions?
+ Who weighs more? smokers or nonsmokers?
+ Are people who regularly wear seat belts less likely to engage in indiscriminate sex?
+ How different are the attitudes of male and female highschool students about gaining weight?

Some answers may surprise you.

A college education is more than just memorizing facts provided in the classroom and spitting them back. This book helps you to prepare to learn how to learn, and fosters an attitude of self-directed lifelong learning through independent exploration and critical thinking.

And that is what a college education should be.

$28.95   ISBN 1–888725-24-9

Heuristic Books

PO Box 7151

Chesterfield MO 63006-7151 USA

Heuristic Books

for Mathematics & Management Science
heuristicbooks.com

Books byT.L Brink

from Science & Humanities Press

Questionnaires--Practical Hints on How to Avoid Mistakes in
Design and Interpretation--By T.L.Brink, PhD, MBA (2004)
Includes bibliographical references and index.   ISBN 1-888725-
74-5 6¼X8¼, 270pp, $18.95

Order form			
Item	Each	Quantity	Amount
Missouri (only) sales tax 6.925%			
Priority Shipping			$5.00
	Total		
Ship to Name:			
Address:			
City State Zip:			

Heuristic Books

for Mathematics & Management Science
heuristicbooks.com

Heuristic Books
PO Box 7151
 Chesterfield, MO 63006-7151
(636) 394-4950
heuristicbooks.com

www.ingramcontent.com/pod-product-compliance
Lightning Source LLC
Chambersburg PA
CBHW081146270326
41930CB00014B/3054